Project Management Methodology

Project Management Methodology

A Practical Guide for the Next Millennium

Ralph L. Kliem

Practical Creative Solutions, Inc.
Redmond, Washington

Irwin S. Ludin

Practical Creative Solutions, Inc.
Redmond, Washington

Ken L. Robertson

KLR Consulting Inc.
Burnaby, British Columbia, Canada

MARCEL DEKKER, INC.   NEW YORK · BASEL · HONG KONG

Library of Congress Cataloging-in-Publication Data

Kliem, Ralph L.
    Project management methodology : a practical guide for the next
millennium / Ralph L. Kliem, Irwin S. Ludin, Ken L. Robertson.
        p. cm.
    Includes index.
    ISBN 0-8247-0088-0 (hardcover : alk. paper)
    1. Project management. 2. Industrial management—Methodology.
I. Ludin, Irwin S. II. Robertson, Ken L. III. Title.
HD69.P75K593   1997
658.4'04—dc21

Marcel Dekker, Inc., and the authors make no warranty with regard to the accompanying software, its accuracy, or its suitability for any purpose other than as described in the preface. This software is licensed solely on an "as is" basis. The only warranty made with respect to the accompanying software is that the diskette medium on which the software is recorded is free of defects. Marcel Dekker, Inc., will replace a diskette found to be defective if such defect is not attributable to misuse by the purchaser or his agent. The defective diskette must be returned within 10 days to: Customer Service, Marcel Dekker, Inc., P.O. Box 5005, Cimarron Road, Monticello, NY 12701, (914) 796-1919.

Comments regarding the software may be addressed to the authors:

    Ralph L. Kliem: 75377.2623@compuserve.com
    Irwin S. Ludin: 75763.1543@compuserve.com
    Ken L. Robertson: ken.robertson@klr.com

The publisher offers discounts on this book when ordered in bulk quantities. For more information, write to Special Sales/Professional Marketing at the address below.

This book is printed on acid-free paper.

Marcel Dekker, Inc.
270 Madison Avenue, New York, New York 10016

Current printing (last digit):
10 9 8 7 6 5 4 3 2 1

PRINTED IN THE UNITED STATES OF AMERICA

To my loving and patient wife, Priscilla
—R.L.K.

To my boys, Mike D. Ludin and Rick S. Cooper
—I.S.L.

To my wife, Judith, for her continual support
—K.L.R.

Preface

The Project Management Methodology has been developed as an international joint venture project by Practical Creative Solutions, Inc., of Redmond, Washington, U.S.A., and KLR Consulting Inc., of Vancouver, British Columbia, Canada. Since it is a practical application, it has taken the nickname P^2M^2, short for *practi*cal *p*roject *m*anagement *m*ethodology.

The methodology is designed for use by project managers in a wide variety of backgrounds. It presents project managers with the fundamentals of how to manage a project. Many inexperienced and some experienced project managers bypass the basics and find themselves faced with tremendous difficulty. P^2M^2 will help to guide project managers through the challenges of managing a project.

The first three chapters of the methodology explain how project management principles have been applied in the past and introduce the basic concepts of the P^2M^2. The remaining chapters cover, in detail, the six processes of the methodology. These processes—leading, defining, planning, organizing, controlling, and closing—are covered in separate chapters. Each process is broken down into activities. Each activity includes an overview and a template that you can use as shown or modify for your particular circumstances.

Templates appear at the end of each chapter. Each template provides the inputs, tasks, responsibilities, and measures of success to execute a process, collectively known as the P^2M^2 cycle. This cycle is iterative, meaning that the cycle repeats itself. Each template may be a subset of a higher process; i.e., the leading (10000) template has five lower level templates: vision (10100), communicating

(10200), motivating (10300), maintaining direction (10400), support (10500), and team building (10600).

P^2M^2 is designed to help project managers. However, the methodology alone will not guarantee you success as a project manager. To be successful, you must put considerable energy into the activities described in the methodology and make these activities a natural part of managing a project. There is no "cookbook" approach to success, but if you use P^2M^2 wisely, it will assist you in your efforts to be a world-class project manager.

Ralph L. Kliem
Irwin S. Ludin
Ken L. Robertson

Contents

1

Introduction: Why Project Management?

Your boss just walked into your office. She pulls a chair up to your desk and hands you a bombshell. She wants you to automate all functions on the entire floor of the building by a specific time and within a designated budget.

You sit still for a minute, more frozen from fear than from the thought of how much effort this project will require. Every conceivable problem, obstacle, and disaster floods your mind. You know the boss won't accept the eternal "It can't be done." The first question that comes to your mind is: What do I do?

Project Management Defined

Project management is the answer to this obvious question. **Project management**, however, is a loaded term that says a lot and means much more. Project management provides you with the tools, knowledge and techniques for leading, defining, planning, organizing, controlling, and closing a project. But what is a project?

A **project** has a defined start and stop date. Every project must begin at a specific point in time and must complete some time in the future. So-called "projects" without end dates are not projects; they are nothing more than an endless continuum of routine work. **Exhibit 1-1** compares the characteristics of a project to those of an operational activity:

Project Activity	Operational Activity
Produces a new specific deliverable	Delivers same product
A defined start and end	Continuous
Multidisciplinary team	Specialized skills
Temporary team	Stable organization
Uniqueness of project	Repetitive and well understood
Work to a plan within defined costs	Work within an annual budget
Canceled if objectives cannot be met	Continual existence almost assured
Finish date and costs more challenging to predict and manage	Annual expenditures calculated based on past experience

PROJECT ACTIVITIES V. OPERATIONAL ACTIVITIES
EXHIBIT 1-1

A project involves actions to deliver a final product. The path to this final product is more than a whimsical orchestration of activities to get from point A to point Z. Certain steps must be performed in a logical sequence, not randomly. To do Step B, you must first do Step A. A project has no room for illogical sequence because it has a series of constraints imposed upon it, such as cost, budget, and schedule.

Projects must deliver a product, or "deliverable," at the end. The product may be a car, a house, a software application program, or a marketing plan; in other words, there must be some tangible result. It should be measurable in some way to verify the quality of the product

Projects Without Project Management

If you must manage a project, you have one of two fundamental options. You can manage the project efficiently and effectively. Or, you can manage your project in a way that results in high turnover, low morale, poor productivity, and ineffectiveness.

Project management helps you to achieve the former. It provides you with the tools and techniques for leading, defining, planning, organizing, closing, and controlling a project both efficiently (resource utilization) and effectively (customer satisfaction).

Think about projects you have worked on, perhaps even managed, that ended either disastrously or successfully. They probably had one of four patterns, which are reflected in **Exhibit 1-2**.

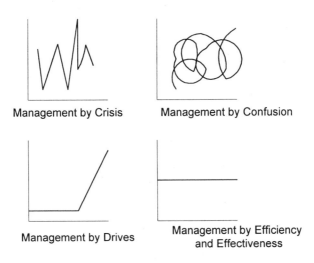

Management by Crisis   Management by Confusion

Management by Drives   Management by Efficiency
                              and Effectiveness

PATTERNS OF PROJECT MANAGEMENT
EXHIBIT 1-2

Under management by crisis project managers find themselves constantly reacting, rather than proacting, to their circumstances. Something happens and they try to fix it. No sooner they had fixed one problem than they have to fix another one. Throughout their project, they are constantly like a military garrison barraged by guerrilla fighters. They face a no-win situation, unable to take the initiative. The best way to describe this operating mode is a stimulus-response syndrome.

Management by confusion is almost as bad as management by crisis. Projects handled in this mode are similar to a person walking around with one foot nailed to the floor; they walk around and around, repeating their steps. Their projects are extremely inefficient and entail much duplication of effort. Little or no progress is the result. Typically, work must be repeated because some slight "oversight" occurs, often resulting in the spending of needless sums of money.

Management by drives is like a person walking through a valley and suddenly climbing a steep mountain. Nothing much happens on the project for a long time. Then, suddenly towards the end, a mad rush to finish the project occurs. Everyone runs about as if having a caffeine fit. Employees work long hours. This not only results in burnout but escalates project costs. The quality of the work suffers. The only item of concentration is meeting the schedule with little regard towards cost and quality standards.

Management by efficiency and effectiveness are well-managed projects. The project managers who follow this mode are in control of the project. Or, if they lose control, they are able to get back into control quickly and effectively. People working on the project know what their responsibilities are and when to perform them. The project managers know what their priorities are; they also know what resources are required to address those priorities. Unlike management by confusion, management by drives, and management by crisis, this pattern has no crooked or bent line; it is smooth.

Two Types of Management Styles

In essence, we are talking about two categories of management styles in approaching projects. Patterns A, B and C are reactive; pattern D is proactive.

A **reactive project management style** has several characteristics. As the word **reactive** implies, project managers subscribing to this management style are constantly behind in reaching the goals and objectives of the project. They tend to be impulsive and excitable, and to have a perspective that is extremely short-range. When making a decision, they think about overcoming the next hurdle, not about whether they are jumping over the right one. They have, therefore, a short-range perspective. This perspective is highly indicative of not being in control. Worse, they rarely plan or organize even themselves. Typically, their projects are behind schedule and exceed budget. Quality is questionable, as is customer satisfaction.

A **proactive management style** has its own unique set of characteristics. As expected, project managers with this style take a long-range perspective. They look at the next hurdle, but they can also see whether they are leaping in the right direction. They are restrained in their decision-making (which is not the same as being indecisive). They weigh all options and then make the right logical choice. They know what is required to regain control, should they lose it. They also plan and organize themselves before proceeding. Typically, their projects finish on time, within budget, and with good quality and happy customers.

You want a proactive style because you seek a project that will end on time and within budget while producing something of quality. To acquire this style, however, requires discipline.

A proactive style involves planning and organizing yourself and your project before proceeding. This entails defining the goals and objectives of the project and its final output, knowing in advance the necessary steps to meet those goals and objectives, and the logical sequence of those steps.

A proactive style also means developing contingency plans in case something goes awry. Contingency plans are strategies to allow the project manager to respond to unexpected situations that may negatively impact the project. The appropriate response is determined in advance to allow the project manager to regain control of the project.

Further, a proactive style entails making a calculated decision, which means foregoing the urge to "shoot from the hip." A proactive project manager knows the value of first inquiring into a problem and then selecting the best solution. This is quite different from the common practice of selecting a solution and imposing it on a given set of circumstances. The latter approach often leads to needless turmoil.

Proactive project managers:

- Determine the major goals and objectives
- Rationally assess the environment
- Develop alternative courses of actions to execute the project
- Select the best alternative
- Implement the selected alternative
- Establish open communication channels and keep them open to continually assess project circumstances and to maintain control

Pressures on Project Managers

Taking a proactive style is not easy. The typical project manager faces a host of problems that can quickly lead to a reactive management style. **Exhibit I-3** shows the types of pressures placed on project managers.

THE PRESSURES AND ISSUES BOMBARDING TODAY'S PROJECT MANAGER
EXHIBIT 1-3

Public Relations. Public relations issues confront project managers. They must be aware of the negative impact a faulty product can have on a customer. If the product continually breaks down or product support (or maintenance) is inadequate, the company's reputation will tarnish. This can translate into no more future business with either current or future customers.

Politics. Politics is everywhere. The project manager must face political pressures from direct management, senior management, suppliers, and customers. One bad move cannot only ruin the project but also destroy the project manager's reputation and credibility.

Economics. Monetary issues always confront project managers; no way exists of avoiding them. What project managers do in the present and in the future depends greatly on available monies.

Goals/Objectives. Frequently, project managers manage a project that must conform to the requirements set by higher management, not just those specified by the customer. Higher management, for instance, may dictate that a project be completed earlier than the agreed-upon end date in the contract to meet a market window. Or, to increase profitability, higher management might approve a budget substantially less than that specified in the contract.

Legal. Project managers may also face legal issues. State/provincial, federal and international laws may require following specific procedures during the project. For example, Department of Defense project managers must comply with specific directives that constrain what project managers can and cannot do.

Personnel. Personnel issues bombard project managers, too. Affirmative action is one common issue in the United States. Project managers may have to acquire resources that fit the appropriate affirmative action mix.

Marketing. Marketing issues also constantly barrage project managers. They must understand the market for their product. They must know what the customer and the end market expect the project to deliver regarding functionality, cost, quality, schedule, and service, both now and in the future.

Business Environment. Along with marketing issues are those concerning the business environment. Project managers must deal with questions like: What are the business conditions today? In the near future? In the distant future? Answers to these questions and others can make a difference in the quality and level of product offered.

Standards. Standards, within the company and outside, pose additional serious issues. Internally, project managers must meet standards set by another part of the company, often known as quality assurance. They may also have to meet standards established by a governmental authority or a corporate customer. Failure to comply with the standards can lead to penalties or cancellation of the project.

Limited Resources. Limited resources are another problem project managers must grapple with in their environment. Few project managers today have access to unlimited resources. Instead, they must optimize the resources they do have.

Technology. Project managers must also confront technological issues. They must know what technology is available to conduct the project. If developing a "hi-tech" product, such as a computer applications, they should know the latest advances in technology that might be applied to the project. In other words, they must recognize the latest advances in technology to prevent themselves from developing an obsolete product that the customer no longer wants or needs.

As you can see, project managers face a host of issues. Unless they establish mechanisms for leading, defining, planning, organizing, closing, and controlling their projects, they can quickly fall into a reactive project management style.

Problems of Reactive Project Managers

Despite the danger of becoming reactive, many project managers fail to institute measures for leading, defining, planning, organizing, closing, and controlling projects. Instead, they repeat the same problems, whether for the same or a different project. A definition for "insanity" is "repeating the same actions, but expecting different results." It just doesn't happen.

Insufficient resources are a frequent problem facing reactive project managers. Poorly managed projects often lack an adequate level of resources to complete successfully. Not having enough people with a specific skill, for example, is a common occurrence. To make matters worse, some project managers wait until the last minute to determine the right number of people with the requisite skills. Consequently, critical activities slide before hiring staff, and then it is usually too late. Often a project is not completed for this reason.

Even if adequate resources exist, some of these same project managers misallocate their resources. For example, they may assign someone to a task that does not utilize their skills or who lack the requisite expertise level. Under such circumstances, the activity usually fails to be completed on time.

Inadequate management of changes is another problem project managers face. No work environment is static; everything is constantly changing, even the requirements established by the customer. Often, project managers fail to manage the changes bombarding them. They consider each change as just as important as the next one. Yet, not all changes have equal importance. Some can seriously impact the project while others lack criticality. Typically, projects without a change control mechanism experience management by confusion or management by crisis patterns. Consequently, the changes manage the project managers, rather than the project managers managing the changes.

Poor communication is another frequent problem reactive project managers confront. If the project managers fails to communicate, then the project resources may not know what tasks they are to perform or even what the major goals and objectives of the project are. Poor communications can quickly lead to frustration, misinterpretations, and negative rumors. People often fail to attend a meeting, for example, because no one has told them about it. People perform tasks that someone else is doing or that are no longer necessary. These are all consequences of poor communications. The source of such problems is simple: these project managers do not communicate to their people.

Low morale is still another problem resulting from project managers using a reactive style of project management. Some typical indicators of low morale are high turnover and absenteeism, which manifest employees' lack of interest in the project. Frequently, employees suffering from low morale tire from their project manager's inability to make consistent decisions or set priorities. Frustrated, they look elsewhere for the right leadership.

Inadequate documentation is yet another problem these managers face. Almost nothing is documented. Documentation is something that "gets done when the time is available." Reactive project managers do not bother to take minutes at meetings, nor do they keep even a simple log of project activities. Worse, they fail to even record everyone's responsibilities. This attitude leads to communication problems and a lack of responsibility among project participants.

Poor documentation leads to other problems, too. It results in duplicate effort and "re-inventing the wheel" because no one knows who is doing what or whether something has been done previously.

Lack of responsibility and accountability is another problem. Many of reactive project managers are reluctant to pinpoint responsibility because that sometimes leads to confrontation between themselves and their people. The preferred mode of operation is to allow people to function autonomously until they complete their work. The results are often territorial fights and disagreements over petty issues. Such circumstances can paralyze the project team. Typically, reactive project managers fail to create an organization chart; instead, they allow for an informal pecking order to arise, which can prove detrimental by allowing power struggles to arise.

Poor planning is common, too. Reactive project managers do not plan even when the time is available to do so. To many, planning is a laborious and tedious process that takes them away from building a product.

Good planning involves deciding in advance what the project will achieve, determining what steps to execute it, who will perform those tasks, and when they must start and finish. Yet, many project managers cannot resist the temptation to begin the project before doing serious planning. They would rather step into the project, ignoring the details. Duplicating effort, repeating work, and allowing poor product quality are frequent consequences of poor planning. Taking time to plan in the beginning allows time in the future to concentrate on critical areas.

Even if they create good plans, reactive project managers fail to track how well their project is progressing. Many project managers develop elaborate plans but quickly discard them once the project starts. To them, plans are nothing more than something to give to upper management and the client. This attitude makes little sense. Why bother to take the time to build plans and then never follow them? Your plans, if done appropriately, are the "road map" for leading you successfully through the project.

Reactive project managers often build plans and follow them until something goes astray. Then they make a decision to replan. While replanning is sometimes necessary, it can lead to rubber-stamping of the project's current circumstances. You should perform such actions sparingly. Your first option is for project managers to do whatever is necessary to come into conformance to their original plan. This leads to better project discipline, something reactive project managers lack.

The Bottom Line on Reactive Project Managers

Projects under reactive project managers lead to negative outcomes. They miss major milestones (events), such as start and stop dates for critical activities or even the start or end of the project itself. Budget estimates are often exceeded beyond what anyone ever expected, forcing these project managers to return to the client or their management for more money. Poor product quality becomes commonplace, which only leads to a tarnished reputation and other, more punitive actions.

The Six Basic Functions of Project Management

No project will be perfect. Problems will beset the best-led, -defined, -planned, -organized, and -controlled project. The key, however, is to anticipate those problems. The best approach is to determine in advance the **who, what, when, where**, and **how** before the project begins. That's called **defining and planning**. It also entails orchestrating your resources with those plans. That's called **organizing**. It involves assessing how well you use your plans and organization to meet project goals and objectives. That's called **controlling and closing**. Embracing the project means motivating people to excel. That's called **leadership**. Leading, defining, planning, organizing, controlling, and closing are the six basic functions of successful project management.

Exhibit 1-4 illustrates the project management process.

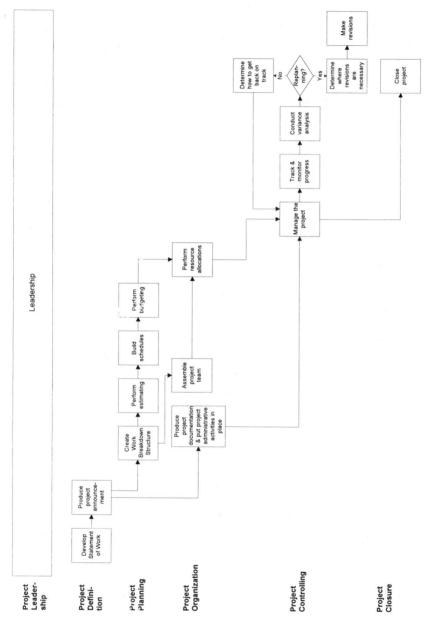

PROJECT MANAGEMENT PROCESS
EXHIBIT 1-4

2

Traditional Project Management Methodologies Versus the Practical Project Management Methodology (P^2M^2)

A growing recognition exists of the impact methodologies have on managing projects. The field of project management is no different from other professional fields. For decades, practitioners of project management have tenaciously adhered to methodologies that appear anachronistic and inadequate; the Practical Project Management Methodology P^2M^2 builds on the strengths of traditional approaches to project management and addresses its inherent weaknesses.

Characteristics of Earlier Methodologies

For years, the field of project management had subscribed to methodologies resembling the rigid, structured, highly rational thinking of business greats (such as Frederick Taylor, Henry Gantt, and Max Weber) for the latter part of the last and most of the current century. These methodologies, referred to as **traditional project management methodologies (TPMM)**, had several characteristics.

1. They assumed the future was predictable. Through estimating, scheduling, and risk analysis, for example, one could forecast with reasonable confidence what will occur and, to a degree, what would not. Schedules reflected a linear, or sequential, approach that anticipated what the future would be.

2. The assumptions about mathematical predictability led to an over-emphasis on mathematics in managing projects. Numbers in project

management became highly important, and mostly functioned as the primary determinant of project success or failure. Just about everything required measurement and calculation.

3. They were highly centralized in their approach. They required building elaborate plans by a few knowledgeable people and having the project manager function as the focal point to plan and execute the project. The project manager, for example, drafted the report after receiving feedback from key individuals who, many times, knew something about the work but did not take any action. In the end, this reflected a proclamation approach to project management, involving minimal participation by most members of the project.

4. TPMM involved viewing projects as a multitude of different functions participating on a project. One function did its "thing" and another its own thing. Somehow the product represented a composite of all those functions. Throughout the project's phases, each function was employed narrowly and set aside, much like players on a football team. Involvement occurred only at certain points in time.

5. They viewed change as an anomaly; something to manage according to prescribed rules. In other words, change was "managed," not "accepted" as the norm. A static model of the project and its environment resided in everybody's mind. Anything deviating from that model was deemed suspect and dealt with accordingly.

6. TPMM emphasized efficiency. It stressed minimizing cost or time. Effectiveness, however, took a back seat, partly because project plans were considered as being built upon achieving goals and objectives. Meeting cost and schedule requirements took top honors for areas of concentration.

7. TPMM focused on cost, schedule, and quality. Often it viewed the relationship among all three as a zero-sum game. If project managers desired good quality, then project managers had to sacrifice cost or schedule, or both.

Results of TPMM

The characteristics of TPMM have often led to an unrealistic way to manage projects. To a large extent, TPMM resembled its long-standing relationship with the construction industry. Outside that industry the principles, tools, and techniques of TPMM have appeared somewhat akin to fitting a square peg into a round hole. In some cases the two fit together; but most often, they do not fit at all well.

Inflexibility. Because it encouraged planning to the smallest detail, TPMM led to a highly inflexible environment, concentrating, for example, on meeting dates that often created an atmosphere of "sticking to your guns," regardless of consequences or other considerations. Subjects such quality took a back seat. TPMM even failed to wrestle well with topics less "rational" or "structured" than itself. For example, TPMM had not adopted too coherently the basics of total quality management (TQM) and business process improvement (BPI). It treated these subjects as an appendix to the project management body of knowledge.

Unresponsiveness. The rational, hierarchical, and structural aspects of TPMM led to an environment where requests falling outside the plan were ignored or not accepted. A static model of the project and its environment resided in everybody's mind. Anything deviating from that model was considered suspicious. If not discarded, it was regarded as an anomaly and treated accordingly.

Customer dissatisfaction. In addition to unresponsiveness to change was the lack of customer satisfaction on projects. This dissatisfaction frequently resulted from an unrealistic desire to comply with the plan to meet measurable goals and objectives. "Getting the job done" on time and within budget became the modus operandi. It was thought that fixes and other revisions could come later on their own or under another budget or schedule, much to the frustration and expense of the customer.

Task-oriented, not results-oriented. The elaborate plans emphasized completing tasks, such as those listed in the work breakdown structure or the schedule. Delivering a quality product was not adequately addressed. Doing the right work was more important than doing the **work right**. Again, it was felt that the fixes and other revisions could come later.

Lack of innovation. The inflexibility and unresponsiveness of TPMM did not create an environment that encouraged innovation. Elaborate plans and pressure to fulfill them encouraged one to comply or else face negative consequences. No time was available for experimenting or seeking alternative paths for project execution. In some cases, creativity was viewed as "play" and not part of running a project. Pressure to fulfill them encouraged one to comply or else face negative consequences.

Extensive documentation. TPMM emphasized extensive documentation. This documentation ranged from compiling a thick project plan to developing elaborate policies and procedures promulgated by the project manager. In the end, piles of documentation filled the project history files.

High overhead. Closely allied with extensive documentation was high overhead. Someone had to develop, compile, and maintain the documentation. In the end, the project devoted considerable staff to administrative tasks. In many cases, increasing the administrative overhead became more important than having the right people to produce and deliver a product.

Overcontrol. TPMM encouraged keeping abreast of the slightest deviations to plans. Tracking and monitoring became so critical that innovation was stifled and an administrative structure arose. Attention to detail was more important than criticality. In the end, mole hills became mountains, and mountains became mole hills.

Overstructuring. Planning in detail involved more than schedules and work breakdown structures. It also involved planning for resources and their utilization. Under TPMM, planning became elaborate to the point that it bordered on the

Theater of the absurd. Planning became so extensive it led to a lack of teamwork by resulting in overspecialization and lack of accountability. In the end, each person became an island.

Overmeasurement. The overwhelming desire to control led to extensive measuring. If you can't measure it, as an old saying goes, then you can't control it. Schedules and budgets were about the only thing that really could be measured with some statistical reliability and validity. This resulted in a fanatical desire to measure. Data, not information, became the norm.

Emphasis on tool or technique, not results. The desire to automate planning resulted in an interesting aberration. Applying and using the tool or technique became more important than managing the project. Great effort was made towards satisfying the needs of the scheduling tool, for example, rather than performing the necessary work. In some cases, the plans changed to satisfy the tool.

More is better. Under TPMM, emphasis was on providing a lot of everything and not necessarily the right amount of anything. Elaborate, but not necessarily relevant, plans became the norm. Analysis paralysis was common; nothing started until the plans were perfect, which often meant irrelevant, unnecessary detail. Even during the execution of the project, more was considered better to acquire an accurate pulse on performance.

Convergent thinking. Finally, TPMM led to extensive convergent, or narrow, thinking. The concentration was on schedules and budgets at the expense of everything else. Often one did not take a divergent, or broad, perspective. Impact on other projects, for example, was rarely considered unless identified early. Essentially, all projects were islands.

TPMM, therefore, appears inappropriate for today's contemporary hi-tech, fast-paced environment. The concepts, tools, and techniques falling under this project management paradigm have necessitated the creation of a new one that strips away the cumbersome, archaic characteristics of TPMM. Hence, the need for P^2M^2.

Characteristics of P^2M^2

The P^2M^2 paradigm has several characteristics that distinguish it from TPMM.

Flexibility. P^2M^2 recognizes that projects occur in a dynamic environment requiring flexibility in approach. Plans are dynamic, malleable instruments for managing a project. They are not to be overwrought with meaningless details (i.e., not everything has to be covered down to the molecular level). In this methodology schedules are not "etched in stone."

Results, not task-oriented. Completing a task on time and within budget are fine, but producing a product of top quality is more important. This quality focus essentially means complying with specifications to satisfy the customer. The customer, not the schedule or the budget, is king. Schedules and budgets should be adjusted to reflect level of quality desired by the client.

More participation. Plans are not the result of a select, wise group of people endowed with special powers. Instead, "democratization" of the planning process is the norm. The days of following "marching orders" are quickly subsiding. Everyone impacted by the project must provide input or feedback on the plans. Failure to acquire "buy-in" can result in nonacceptance or noncompliance that impacts the likelihood of success.

Greater accountability. Each participant has responsibility for a unit of work. Each team member is responsible for the delivery of quality output. This responsibility encompasses planning, organizing, and controlling for its success. Each participant does not perform a task per se. Instead, each performs tasks that result in completing a unit of work. In other words, a person's work results in output with that person's "signature" on it. Doing a portion of a task only to let someone else handle the rest is unacceptable.

Minimal hierarchy. Organization charts and responsibility matrices are not more important than teaming to complete a unit of work. Stress is on keeping the lines of communication open and not imposing layers of hierarchy. Adding more people and layers of supervision do not expedite a project but only encumber it. Less is better. Emphasis is placed on self-management.

Less documentation. P^2M^2 views plans of great detail as unnecessary. Completing countless forms to collect data and assess schedule and budget status is not only wasteful but irrelevant to delivering a product of high quality. The idea is to have the documentation capture the right data to produce the right information at the right time at the right place. Too much documentation means too little information.

Results more important than tools or techniques. Stress is on obtaining results efficiently and effectively, not complying with the requirements of an automated scheduling tool or elaborate plans. In other words, the "tail does not wag the dog." If the tool or technique does not expedite the execution of plans, the tool or technique must go. The concentration is on developing a product of high quality.

Right measures. The idea is not to measure for the sake of measuring or to collect statistics on everything; it's about developing the most appropriate measurement and then turning data into information. The goal is to collect the right data at the right time and turn it into the right information, at the right moment, for the right people. Again, less is better.

Divergent thinking. Concentrating on the total picture means taking a broad, cross-functional and, just as importantly, organization-wide viewpoint. Each project consists of a set of functions and processes that must work together smoothly to deliver a product. Integration is key. Integration entails looking at a project from a grander scale to see its impact on the organization as a whole.

Continuous improvement. Under P^2M^2, the project environment is dynamic and ever-changing (i.e., not static). Plans must not be viewed as "etched in stone" but as living approaches for completing projects. Change is accepted and not merely managed. Change contributes to the progress of a project rather than being an obstacle to overcome.

Integrated teams. P^2M^2 views projects as having multiple disciplines cooperating to provide a product of high quality. Projects are not viewed as consisting of an elite corps of highly specialized functions with other ones as mere appendages. Each discipline has a significant role in contributing to the successful completion of the project. No room exists on a project for an oligarchy of disciplines.

More awareness of irrational processes. P^2M^2 recognizes the importance of the less rational factors of project management. Not everything on a project occurs in a straight line. Instead, a project consists of spirals and crooked lines that often make no sense at all. Under P^2M^2, a project needs people with skills that can handle such circumstances. A project has just as much propensity towards disorder as order. All pieces need not fit nicely together nor should they. The "fog of project management" is a natural state. Skills related to interpersonal relations and creative problem definition and solution become as relevant and just as important as, for example, engineering or programming.

Keystones of P^2M^2

P^2M^2 is more than a set of principles, tools, and techniques. It is a philosophy as well as a methodology for completing projects that deliver a product that satisfies the customer. Several keystones define the practical project management methodology.

Keystone No. 1: Customer Satisfaction

P²M² stresses customer satisfaction above all else. Project managers need to meet the customer's requirement even if it means sliding the schedule or exceeding budgetary limits (assuming the customer agrees with exceeding the schedule and budgetary milestones).

Making customer satisfaction the driving force for a project is not new. Business process reengineering and total quality management do the same with success. Project management is a vehicle for doing the same. The customer is the reason for the project.

Keystone No. 2: Balance

If the customer doesn't agree with sliding the schedule and budgetary limits and insists on higher quality, then finding a judicious balance between cost, schedule, and customer satisfaction become important. "If push comes to shove," however, customer satisfaction has preference, and schedule and budget must be adjusted accordingly. In the end, the customer receives what the customer is willing to pay for.

Customer satisfaction does not occur in a vacuum. It requires money and time. The level of satisfaction will be dependent, to a major degree, on the amount of time and money the customer is willing to expend. Under P²M², all three are interrelated.

Keystone No. 3: Participation

Project plans reflect the wide participation of all participants, not a select group of individuals. The people executing a task should have a direct say in the time and resources required for its completion. They are most familiar with what must be done and have a stake in the outcome. This "democratization" of the planning process engenders greater ownership and, consequently, responsibility and accountability.

The level of participation depends on the extent a person or organization is impacted by a project. The greater the impact, the higher the level of participation should be. This level of participation should occur as early as possible and be sustained throughout the life of a project.

Keystone No. 4: Self-management

P²M² stresses the need for team members to assume greater responsibility to complete tasks and products ("deliverables"). This cornerstone ties closely with the previous one. With greater participation comes more responsibility for successful execution. Team members are held accountable for planning, organizing, and controlling all activities associated with the activity assigned to them. Responsibility for success or failure rests with the team members responsible for its completion.

Unit of work is the principal force behind self-management. This entails having individuals responsible for producing a finite, tangible deliverable as a result of their labors. They plan, organize, and control the development and implementation of that deliverable.

Keystone No. 5: Modularity

Modularity entails exploding a project into manageable components. Under P^2M^2 it becomes the basis for reusability for project plans. The reusability of modules becomes extremely critical in constructing work breakdown structures and network diagrams. These modules avoid unnecessary task replication and shorten the planning cycle for future projects.

A key ingredient in achieving modularity is establishing a library function, which is responsible for cataloguing and maintaining all project plans and identifying opportunities for reusing components of those plans.

Keystone No. 6: Process improvement and reengineering

Under P^2M^2, project management tools, such as schedules, are as much tools for process improvement and reengineering as they are for time management. This cornerstone is allied to the last one because of the importance of simplicity. When reviewing schedules, project managers and all team members should strive to minimize delays, inspections, approvals, duplicate tasks, setup times, and reviews. Work breakdown structures and schedules, therefore, become more than a laundry list of tasks to complete at specific points in time.

Project plans provide ample opportunity to analyze approaches for completing projects. A key strategy is to identify opportunities not only to shorten time but reduce inefficiencies. While saving time is one way to reduce cost, it is not the only way. Project plans enable identifying those other opportunities.

Keystone No. 7: Non-linearity

TPMM viewed projects as being linear, flowing in some predictable pattern. The reality is that this pattern is spiral rather than sequential or concurrent. The spiral nature of projects stems from the need to return to previous tasks to refine earlier work.

Under TPMM, project management tools did not reflect this re-visitation of tasks. Under P^2M^2, the notion of returning to a previous task is a natural occurrence. Circumstances such as disapprovals, incompleteness, and oversights can and will occur during the normal course of events. Acceptance of that reality enables realistic planning.

Keystone No. 8: Metrics

P^2M^2 stresses building in quality during the project. Quality is not some afterthought. Implementing quality requires establishing metrics to track and assess quality. These metrics must measure the right processes. It's not just a matter of measuring but having the **right** measures to measure the right things.

Project plans should reduce the number of inspections and other opportunities for rework. The best approach is to implement "best practices" measures within the project.

Keystone No. 9: Integration

Under TPMM, the project team included an elite core of highly specialized individuals who called the shots. This view produced partial results. Under P^2M^2, a project team is viewed as an integrated, multifunctional entity to deliver a product yielding complete results. Participation by everyone is essential to ensure not only the completion of a project but also its completeness.

To encourage and maintain this integration, all team members are ideally located in one area rather than being dispersed across a wide geographical area. In addition, these team members work together for the entire production of a unit of work and, preferably, during the entire project.

Keystone No. 10: Simplicity

TPMM complicated the project management process by creating elaborate plans and establishing an administrative machinery. This obscured major issues and turned minor ones into bigger concerns. P^2M^2 stresses simplicity to focus more on major issues and less on the minor ones.

The key for achieving simplicity is buy-in from all those individuals or organizations impacted by the project. Quite often, what distinguishes the major from the minor issues is what project participants view as important. Once they provide guidance on criticality, the project team can then direct its efforts accordingly.

Another way to achieve simplicity is to develop a set of standards and heuristics for planning. Examples include restricting the number of tasks in work breakdown structures (WBS), adopting a common estimating technique, and requiring certain tasks in every WBS and schedule.

Keystone No. 11: Motivation

Motivation is essential for having the project team perform at its highest capability. Motivation through leadership, not management, is an integral part of project success.

A mechanism for achieving motivation is to empower team members by granting them the authority and latitude to participate not only in the decision-making process but also in the ability to execute decisions. Ownership grows and, consequently, so does a sense of responsibility which, in turn, means greater motivation.

Keystone No. 12: Flatness

With greater participation and self-management comes less hierarchy. Under P^2M^2, the fewer layers of management within the team it must deal, with the better. The team can then concentrate on building the product, not hassling with administrative overhead. This, in turn, means fewer delays due to reviews and approvals.

Flatness means greater span of control. Project managers must be willing to relinquish some power to flatten the hierarchy and increase motivation. In addition, the project manager and key team members must be willing to absorb administrative and managerial responsibilities previously assumed by the now missing levels.

Keystone No. 13: Adaptability

P^2M^2 recognizes the project environment as a dynamic, ever-changing one. Adapting to such an environment leads to greater survivability of the project. Inflexible planning and other management practices can result quite easily in an unrealistic, even a surrealistic, approach to managing projects.

Under P^2M^2, replanning is not viewed as weak or indecisive. Rather, it is a sign of management being attuned to its environment and being able to make realistic appraisals about the course of the project. This does not imply abandoning the vision of the project, only being prepared to develop alternative approaches to achieving the vision.

Keystone No. 14: Divergence

Under TPMM, the tendency of project managers was to have a narrow focus. Under P^2M^2, the focus broadens such that a project is looked at as a system involving a wide range of players and disciplines working together harmoniously to meet or exceed customers' needs. This does not however, mean, overlooking the main issues of a project. Nor does it mean looking at the total picture and then identifying the main issues.

One way to achieve divergence is to have project managers who may not necessarily have technical proficiency but knowledge about and exposure to a wide range of issues.

Keystone No. 15: Effectiveness

Efficiency does not always equate with effectiveness. **Efficiency** deals with performing activities with minimal waste; **effectiveness** addresses whether the project is achieving its goals and objectives. The driving force of effectiveness is customer satisfaction. Under TPMM, the stress was on efficiency, with little thought given to effectiveness. Under P^2M^2, the stress is on effectiveness. This requires greater patience by management, which often seeks immediate results for minimum effort. An effective project, however, can reap much greater rewards in the long run than by saving a few pennies at the moment.

The idea is to focus on the vision of what the project must achieve. All effort is then directed towards making the vision a reality. One way to determine and assess progress in attaining that vision is to establish benchmarks or measurable objectives validated by metrics.

Need for Change

To a large degree, TPMM had its origins in the construction industry. Project management survived well in such a predictable, repeatable environment that has enjoyed a long history. In most other industries, the business environment is not as predictable, making it difficult to apply TPMM and other older project management techniques in a world that is dynamic and ever-changing. The key to successful project management in today's world is P^2M^2.

3

The Practical Project Management Methodology (P^2M^2)

Overview

Although the disciplines of project management have existed in one form or another for several decades, very few project managers have implemented them, let alone know what they are. Some even view project management as simply developing schedules. Such perceptions are erroneous.

Project management, like other disciplines, has evolved to encompass a wide spectrum of topics. Today, it still faces many obstacles. Yet these obstacles are not insurmountable. Project managers can—and do—overcome them.

A Short History

In the early 1900s, Henry Gantt, a famous industrial engineer, heralded the beginning of current project management practices by developing the famous Gantt chart, or bar chart (which will be covered in detail in Chapter 6). The chart was a simple way to show the number of days a specific task took to complete.

Later, advancements in production planning and control sparked the rise of several project management practices. Then, in the late 1950s, two important planning techniques arose. These were known as **PERT (program evaluation and review technique)** and **CPM (critical path method)**. These techniques enable project managers to develop elaborate plans for projects in both the public and private sectors.

Yet it would be a mistake to assume that project management is simply a matter of building schedules. Project management involves more than merely scheduling, despite its criticality. Project managers must also organize and control their projects. Appropriate structuring of project resources assists the project manager in turning plans into reality, and appropriate controls help to ensure conformance to these plans.

Project management has been around for a long time—well before Henry Gantt. Generals throughout the history of mankind have used project management. The victorious generals have known the importance of having a good plan, which included knowing what the mission is and the requirements to accomplish it. They organized their resources in a way to minimize inefficiency and maximize effectiveness. They also knew good plans and organizations are not enough. They needed to have the right level of control to become victorious. Working according to plans as much as possible in order to maintain control does not mean denying the need for flexibility. A victorious general knows that sometimes replanning is a sometimes necessary.

Generals are not the only ones who have used project management. The pharaohs had to use some form of project management to build the pyramids. They developed elaborate plans before proceeding. Could you imagine hauling a block of stone weighing tons and putting it on the wrong spot and then having to move it to the correct location while in the hot sun? One hopes the pharaohs employed elaborate planning. They also knew what resources and organizations were required to build the pyramids, and presumably they knew how to ensure that actions conformed with plans.

Of course, you are probably not a general and certainly not a pharaoh! Your requirements for how elaborate you want to lead, define, plan, organize, control, and close your project probably vary widely from those of other project managers. This leads to a very important question: When do you use project management?

Six Basic Functions of the Practical Project Management Methodology

Effective project management requires more than developing schedules. It entails performing six basic functions. These six functions are described below.

1. Leading. Project managers must lead. They must create an environment that encourages the best in people and in a way that meets project goals and objectives.

Project managers do not lead only during certain times of a project; they must provide leadership throughout the entire project. Leadership is multifaceted, comprising a wide range of areas. Some areas are discussed below.

Motivation. Effective project managers recognize that project management involves more than building good plans, establishing an efficient organization, and maintaining effective controls. They know that people involved with the project must want to follow the plans, be part of the organizational structure, and abide by existing controls.

Project managers must motivate everyone on the project, not just team members. They must motivate their customers to provide information necessary to deliver a useful product. And they must motivate management to provide the support—be it money, time, or talent—to complete the project successfully. Project managers consequently serve as a catalyst for getting people to participate effectively. They are the ones who ensure that communication occurs among all parties and that the communication serves the good of the project.

Project managers also engender teamwork among all project participants. They act as a catalyst to get the customer, project team, and management to work together harmoniously to meet project goals and objectives.

But project managers have an additional responsibility. They must set an example by establishing performance standards to follow and must expect others to do likewise. They expect much from the project participants but should require no less from themselves.

Traditionally, leadership has been overlooked by many professionals in project management, who have placed emphasis on schedule, quality, and cost as the three most important variables to consider when managing a project. Today, many project managers recognize that people are integral variables in the equation. Ironically, a project can fail if too little attention is placed on managing people. The results are often high turnover, low morale, and less productivity. People are just as important as quality, schedules, and budgets as far as making a project a success.

This lack of appreciation is due partly to uncontrollable pressures. Management or the customer may dictate that segments or all of the project be finished by a certain time or within a specified level of quality, regardless of the impact on the people, thereby, comprising consideration of the project team members. Yet if people are not considered a crucial element, the project will fail, even in the presence of good plans, organizational structure, and proper controls.

The traditional triangle of project success comprises three variables: schedule, quality, and cost. The new paradigm triangle, however, includes people, who remain the focus of all activities (see **Exhibit 3.0-1**).

2. Defining. This function, although short in time, has a tremendous impact on how well a project progresses towards successful completion. It involves defining at a high level the **who, what, when, where,** and **how** of a project.

An important task is to define the major goals and objectives of the project. Project managers must know specifically what the final product will be. A jetliner?

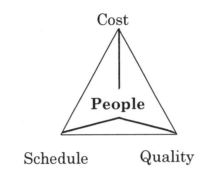

Cost

People

Schedule   Quality

THE FOUR VARIABLES OF PROJECT SUCCESS
EXHIBIT 3.0-1

An automobile? A computing system? But, as a project manager, you must do more. You need to determine specific objectives to reach. If building an automobile, you need to define the criteria it must meet, such as adhering to specified minimum levels of quality. Your objectives should be specific and measurable, and they should serve as benchmarks to determine how well the project progresses.

3. Planning. The project manager and his or her team will determine, to the best of their abilities, a series of actions to complete the project. To accomplish this major feat, the project managers must initiate a planning process.

As a project manager, you must determine the tasks required to reach the goals and objectives documented in the definition component. A definitive listing of tasks provides you with the necessary information to anticipate the quantity and quality of resources to complete the project. It also enables you to make meaningful time and budget estimates. Finally, it gives you a good "stick" against which to measure the degree of progress made on your project. You can present each person on the project with a listing of specific tasks and have them provide you with feedback at regular intervals.

With task listings, project managers can create valid schedules. Each task is tied to other tasks to show the sequence of action throughout the project. In addition, each task receives a time estimate for completion. Project managers can compile this information to determine the start and end dates for each task. Such planning allows project managers to know exactly what tasks must occur and when. A good schedule is a road map through the complex world of a project.

Project managers must also determine which tasks are significant and which are not. Lacking such knowledge can result in concentrating on the wrong tasks and may lead to considerable rework, which can result in needless costs and lower productivity. Effective planning requires identifying the most critical tasks to avoid such problems.

To complete their projects, project managers must identify the necessary resource requirements. They must also determine what is the optimal level of resources to complete their project. Using too many resources results in waste and excess operating costs; using too few can cause negative competition among employees and delay completion of critical tasks. To complete their projects, project managers must know in advance not only what resources are available but also the right amount.

Finally—and related to the previous point—project managers must determine how much money to spend on each task and on the entire project. Having accurate cost information can help the project manager and the line manager to develop a baseline to measure how well resources are being used.

While you probably agree with the need to plan, you know that planning does not come easily. More often than not, you face a number of obstacles.

Perhaps the most common obstacle is lack of time. Many project managers want to plan but find it takes too much time. The irony of the situation is that taking time up-front to plan makes more time available later in the project. An effective way to overcome this perceived obstacle is to prioritize your time to give yourself time to plan or to get some help.

Lacking good planning skills is another obstacle. Some project managers do not plan because they have never learned how to do it. Two effective ways to overcome this problem are to attend training in project management and to get the help of someone who is trained in planning.

Some project managers need more information to plan. They lack the details necessary to prepare useful, definitive plans. Rather than develop a plan based upon assumptions, they become paralyzed. They do not plan, partly because they are fearful of committing themselves to something without having all the facts. While this shows healthy caution, carried to an extreme it can lead to extreme indecisiveness. If you must develop plans based on insufficient information, you can minimize the negative consequences by documenting your assumptions and sharing these with the project sponsor.

Lack of cooperation can hinder efforts to build good plans. You may lack cooperation from the customer, senior management, or even the project team. If some or all of these parties refuse to cooperate, you will lack the information necessary to build good information. One way to overcome this problem is to document instances of poor cooperation and send a copy to the project sponsor. Then ask your sponsor to contact the appropriate people to obtain the cooperation.

Project managers may also face an inadequate level of resources (e.g., people, equipment, material, or supplies). Effective planning is not cheap. You must have the right resources to do the job; people without the requisite planning skills can develop inaccurate, useless schedules. Equipment lacking the right capabilities can also slow the planning process.

Finally, lack of desire on the part of the project managers can kill any chance of effective planning. Planning involves extensive work, which some project managers hate to contemplate, claiming that it takes time away from "doing the project." Yet planning is critical to doing the project right the first time. About the only solution to this problem is for management to make it mandatory to follow a planning process. Senior management should assign project managers before a project starts.

4. Organization. The project manager must institute a structure to maximize the efficiency and effectiveness of the project. Developing a good communications structure for the project is a prerequisite for success. This involves employing mechanisms for effective communications up and down the chain of command. Too frequently, project managers believe they only need to develop good plans and announce them; they assume (incorrectly) that the information will get to the people who need it.

Such an attitude is misleading, even dangerous. Project managers must work at building good communications, both with their team and with senior management. Project managers must establish mechanisms to facilitate effective communications.

There exists a corollary to this last point. Project managers must know when to communicate with other project managers. This is especially important when it is likely that the project will not be completed on time or if it appears that the product will not be delivered to the customer on time or at the level of quality specified. Hence, good communications must occur horizontally as well as vertically.

To have effective communications, project managers must hold the right meetings at the right time with the right people. They must have forms in place to capture only the essential elements of information and do not create a "bureaucratic" atmosphere that detracts from the performance of work. They must develop and implement ways that will facilitate the access and distribution of project information, such as reports and files. Finally, they must establish polices and procedures that everyone not only understands but follows.

5. Controlling. Having good plans and an effective organization is not enough. Project managers must ensure that their projects proceed according to their plans and in a manner that maximizes the effectiveness of their organizational structure.

Controlling involves receiving feedback on the status of a project and, if necessary, taking whatever appropriate action to regain control. Maintaining effective control involves receiving information in three areas: schedule, budget, and quality.

Regarding the schedule, project managers must perform tracking and monitoring to assess whether progress is being made over time. Tracking involves collecting information about what has been done on the project to determine its progress. Monitoring entails looking at not just the past but also the future. The project manager can use this information to predict whether the project will be on schedule, if current trends continue.

Project managers must track and monitor costs. They must know the amount spent and anticipate future expenditures if current trends continue. Like time, money is limited. Usually the problem is too little rather than too much. Project managers must, therefore, account for every dollar spent and ascertain how much they will need to complete.

Project managers must also track and monitor the quality of the output from their projects by having an idea of how many "discrepancies" or defects exist and are permissible. To track and monitor quality, project managers must install measures to ensure that the quality of the project team's output is maintained. Holding product review meetings and inspection reviews are two tools project managers can employ to track and monitor quality of output.

Whether tracking or monitoring schedules, budget, or quality, project managers must have an effective feedback loop in place. This feedback loop should encourage reporting of information—good or bad—up and down the chain of command. Reporting only good information may give a false impression that no problems exist; reporting only negative information may also leave an inaccurate impression. You should seek a balance between the two extremes in order to make intelligent decisions. Avoid falling into the trap of listening to only what you want to hear.

Project managers must know how to detect variances regarding schedule, cost, and quality. They must be able to determine when the actual behavior of their project has gone awry; that is, when deviation from their plans occurs. Ideally, project managers try to adjust actual performance to conform to their plans.

Not all variances are good or bad. A project may be ahead of schedule, indicating perhaps that the quality of workmanship is less than satisfactory. On the other hand, being ahead of schedule may be a positive indication if all the work has been completed satisfactorily.

A variance is a clue that something has occurred and needs attention, perhaps immediately. Project managers must respond quickly, identifying the source of the variance and make a judgment concerning it. If they interpret a variance as negative, project managers must respond quickly and effectively to eliminate it. That means doing whatever is necessary to alter actual circumstances—whether the action affects schedule, budget, or quality—to match their plans.

Periodically, project managers may have to replan. They must determine whether the original plan is no longer relevant. This circumstance may arise when unforeseen difficulties surface, such as failure to get the component of a product to work; inaccurate time estimates for tasks; or senior management's arbitrary reductions to your budget.

Replanning should occur sparingly. Too much replanning indicates having little control over the project and may be construed as conveniently "rubber-stamping" your current circumstances. There is a difference between being **flexible** versus **fluid**. Flexible means adapting to your circumstances but with discipline; primary emphasis is placed on managing a project so that actual circumstances conform to plans. Fluid means having no discipline; that is, accommodating your current circumstances rather than exercising self-discipline. Being the former, you exercise at least some control; Being the latter, you exercise none.

6. Closing. Closing is often treated nonchalantly by amateur project managers. It can mean the difference between a graceful, smooth project completion (or phase completion) and a wasteful, bothersome one.

For this function, project managers compile data and convert it into information to assess performance and provide lessons for future projects (or phases) of a similar nature. They also document their experiences and wisdom to enable future project managers to learn from their experiences—not only the negative but the positive—such as best practices. Finally, project managers smooth the transition from a project environment to an operational one.

Major Players

The major players who participate in most projects include the project manager, project sponsor, project team, client, and senior management. Each one plays an essential role in completing projects successfully.

Project manager. This individual is responsible and accountable for the outcome of a project. The project manager becomes the focal point of the project. Project managers take a leading role in defining goals and objectives, developing project plans, ensuring their projects are executed efficiently and effectively, and maintaining vigilance over progress.

Project sponsor. This is the individual to whom project managers report and it is the person who makes most major decisions regarding the fate of projects. Project sponsors are often responsible for appointing the project manager, securing funding for the project, facilitating the acquisition of resources, approving key deliverables, and providing overall guidance to the project manager.

Project team. The team supports the project manager by working together efficiently and effectively in order to deliver a product that satisfies the client's needs. The team consists of a mix of skills and talents that complement one another.

Client. This is the person (or organization) for whom the product is being built—either for itself or on behalf of another.   The client communicates requirements, dedicates resources (principally personnel) to help the project team, approves deliverables, and is the ultimate user of the project's primary deliverables.

Senior management. Senior management is responsible for determining which projects will be initiated and for assigning a project sponsor to each project. Senior management is also involved at key points throughout the project to receive information updates and to participate in decisions that have an impact on the corporation as a whole.

Ad hoc. Other groups such as project steering committees, advisory groups, quality assurance, and so forth may be involved with the project at the discretion of the project sponsor, project manager, or senior management.

Exhibit 3.0-2 shows the project manager as the common thread among project participants. The project manager interacts with all these people/groups and must ensure that they are all pulling together towards the successful completion of the project.

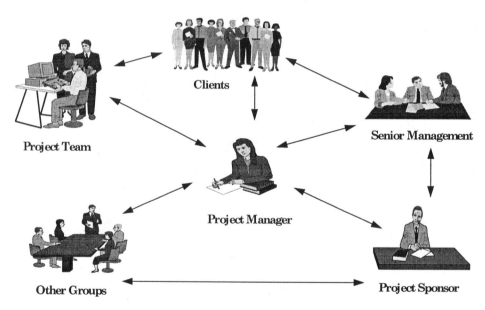

RELATIONSHIPS AMONG PROJECT PARTICIPANTS
EXHIBIT 3.0-2

Art Form, Not Algorithm

It is probably apparent to you by now that project management is not an algorithm. It isn't something you just wind up and place in your work environment, hoping that everything turns out all right.

Employing project management is an art. It requires knowing under what circumstances to use it and what aspects of it to use. No algorithm exists for making that determination. Only you, the project manager, can make these judgments. Under certain circumstances, you may deem it appropriate to develop general plans, while under other situations you may elect to create detailed ones. You may elect to control only certain aspects of a project or manage it in its entirety. The decisions are entirely up to you. That is why you are the project manager.

When to Use Project Management

Project management is not a series of linear activities. To be successful, you must constantly evolve through the six basic functions. The project manager is like a juggler, trying to keep all the balls in the air at one time. The leading function becomes the glue of project management.

Although no hard rules exist, business ventures involving the development of a well-defined product, possessing an independent budget, and having designated start and completion times are excellent candidates for instituting project management practices, regardless of the industry. Typically, project management is seen as something used only in engineering, information systems, and construction. That is a mistaken notion. You can use project management disciplines for business ventures related to such diverse areas as marketing, finance, human resources, pharmaceuticals, forestry, and recreation.

On the other hand, you should not consider project management for activities that are ongoing efforts, that is, business ventures typically lacking start and stop completion dates and a well-defined result. Information systems projects in the sustaining mode are prime examples. Similarly, a service industry might not have a strong need for project management. The effort is ongoing and there is no definable product resulting from the project. Problems naturally arise in such activities, but the problem-solving effort does not involve elaborate planning or follow-through. However, if a specific problem arises requiring considerable planning and follow-up, project management is a definite candidate for use.

Introduce Project Management Early

As you can see, project management may not always be feasible, or even necessary. When project management **is** necessary, however, you should institute it as early as possible. Avoid the circumstance of madly developing plans, organizing resources, and instituting controls after the project begins. It is best to be proactive rather than reactive. There are several reasons for this.

Avoid developing plans to "rubber-stamp" what has already been done. This can misleadingly make a bad project look good. Second—and ironically— the urge to institute project management after a project begins is a confession that the project has not been proceeding very smoothly. Otherwise, why the sudden rush to develop plans? Finally, the benefits from introducing project management disciplines late will probably be marginal at best. Most people on the project will be preoccupied with other activities demanding more of their attention; they will view project management as a nuisance and administrative burden rather than as an asset.

Obstacles to Project Management

Introducing project management disciplines is not an easy task. Some people resist project management practices because they feel it impinges upon their "professional independence." They see it as a way of treating them as children, keeping track of their work.

Others resist project management because it smacks of a police state mentality. They see it as management tracing every move, waiting for the moment to discipline the employee for the slightest mistake, such as missing a completion date or violating a procedure.

Still others fight project management because they feel it inhibits creativity. They claim that working to a timetable makes employees feel rushed and that, consequently, such an environment represses the urge to create. Ironically, the program evaluation review technique (PERT) and other project management techniques originated in research and development environments—places that foster creativity.

Some people resist project management practices for no other reason than perceived additional administrative hassles. They do not realize the benefits of building sound plans, instituting a solid organizational apparatus, and providing good controls and believe that project management disciplines are nothing more than measures for interfering with people doing the "real work."

While under rare circumstances such objections may have some credence, overall they have little validity. Project management does not inhibit creativity; it directs it. Creative writers do not just put sentences together haphazardly to create an essay; they follow the rules and organization of the language to produce a comprehensible output.

Nor does the use of project management smack of police state mentality. If management **does** misuse it as a weapon, then the fault lies with management, not with project management. Project management is a tool, just like management by objectives (MBO) and quality control, to regulate work performance. Its abuse, like other methods of assessment and evaluation, depends on who is using it.

Nor is project management an administrative hassle. Management needs it to have some reasonable assurance that people know how to reach their destination and keep track of how well they are proceeding towards it. Quite often, people who see project management as a hassle are fearful that it may disclose something unsatisfactory about their performance.

The Practical Project Management Methodology: The Answer

Project management comprises six basic functions: leading, defining, planning, organizing, controlling, and closing. The project manager must juggle changing priorities of each to lead the team through these basic functions.

Leading

If one of the juggler's balls falls, the entire project may collapse. Missed schedule dates, cost overruns, and poor workmanship are an indication of more "dropped balls"—and in all likelihood an unsuccessful project. But effective project management ensures that the balls never fall, or that if they do, they can be picked up quickly and brought back into play so that the process can continue.

P²M² OVERVIEW

P² M² CYCLE

INPUTS — TASKS — RESPONSIBILITIES — OUTPUTS — MEASURES OF SUCCESS

INPUTS			**P² M² CYCLE**	**RESPONSIBILITIES**														
Project Management Principles and Techniques	Leadership Principles	Operating Environment		Project Manager	Senior Management	Project Sponsor	Project Team	Client	Cost Calculations	Resource Analysis	Risk Control	Time Estimates	Work Breakdown Structure	Project Goals and Objectives	Project Announcement	Statement of Work	Teambuilding	Support
			TASKS															
		■	00000 Project Environment Assessment	■														
	■		10000 Project Leadership	■	■	■	■	■									■	■
■			20000 Project Definition	■	■	■	■	■						■	■	■		
■			30000 Project Planning	■	■	■	■	■	■	■	■	■	■					
■			40000 Perform replanning	■	■	■	■	■										
■			50000 Project Control	■		■	■	■										
■			60000 Take corrective action	■	■	■	■	■										
			MEASURES OF SUCCESS															
			Has the project environment been adequately determined?															
			Is the project getting positive leadership?														■	■
			Is a meaningful project definition in place?											■	■	■		
			Do useful project plans exist?						■	■	■	■	■					
			Does a project infrastructure exist?															
			Is the project achieving its goals and objectives?															
			Is the project ending cost-effective?															

OUTPUTS

Direction	Motivation	Communications	Vision	Environmental Assessment	Contingency Planning	Status Collection and Assessment	Project Manual	Project Office	Newsletters	Memos	Project Library	Reports	Forms	Workflows	History Files	Procedures	Team Organization	Automated Project Management	Schedules	Winding Down Activities	Post Implementation Review	Lessons Learned	Statistics	Connective Actions	Replanning	Change Control	Meetings
				■																							
■	■	■	■																								
																		■									
						■	■	■	■	■	■	■	■	■	■	■	■	■									
					■	■																		■	■	■	■
																			■	■	■	■					
				■																							
■	■	■	■																								
																		■									
						■	■	■	■	■	■	■	■	■	■	■	■	■									
					■	■																		■	■	■	■
																			■	■	■	■					

00000 PROJECT ENVIRONMENT ASSESSMENT

P² M² CYCLE

History of Projects	Policies and Procedures	Major Participants	Pressures	Management Approaches to Projects	Common Project Problems	TASKS	Project Manager	Senior Management	Sponsor	Project Team	Client	Project Management Assessment	Project Management History
				■	■	**00005** Determine which projects in your environment are examples of management by crisis, drives,	■					■	
			■			**00010** List the major pressures confronting projects in general and project managers in particular in your company	■					■	
				■		**00015** Determine whether your company generally follows a proactive or reactive approach towards managing projects	■					■	
					■	**00020** List some major problems confronting project managers in your company	■					■	
				■		**00025** Assess how effectively project managers are performing the four fundamentals of project management	■					■	
		■				**00030** List the major participants on your project	■					■	
	■					**00035** List the major internal policies and procedures that affect your project	■					■	
		■				**00040** Determine the external actors that you must deal with	■					■	
			■			**00045** List the forces or circumstances that could cause disequilibrium	■					■	
■						**00050** Develop a historical overview of project management in your company	■						■
				■		**00055** Determine what environments within your company could best use project management knowledge, tools, and techniques	■					■	
	■	■	■	■	■	**00060** List the obstacles you or your management could face when implementing project management	■					■	
■	■				■	**00065** Determine ways to overcome these obstacles	■					■	
						MEASURES OF SUCCESS							
						Has an assessment of the company's history of project management been						■	■
						Has an assessment of the project's environment been done?						■	
						Are all the key players identified?						■	
						Are the project management practices the company does well identified?						■	
						Are the project management practices the company does poorly identified?						■	
						Are the policies, procedures, practices, etc. affecting project management?						■	

4

Project Leadership (10000)

Introduction

Leadership, in general, and project leadership, in particular, are an important function in the successful completion of any endeavor. Regardless of the size or complexity of a project, leadership can have a tremendous impact on the results achieved. But what exactly is leadership?

Leadership

Not everyone agrees on what constitutes leadership. Most experts can't even agree on whether leadership is inherited or acquired through experience, or whether it is a science or an art. But they all agree on one thing: that it has a tremendous impact on the outcome of a project in terms of cost, schedule, and quality.

A common thread in the definitions of leadership is the concept of influencing people in the attainment of an organization's goals and objectives. From a project management standpoint, project leadership can be defined as influencing people to achieve the goals and objectives of the project.

In order to influence people, project leaders must accept certain responsibilities. These go beyond managing the project to inspiring people to perform effectively and efficiently. These responsibilities are

> Providing vision
> Communicating

Motivating
Maintaining direction
Being supportive
Team building

A description of each of these leadership responsibilities is found in Sections 4.1–4.6.

Qualities of Project Managers

Being a project manager is not easy. Project managers must have or develop qualities that will allow them to influence people to achieve project goals and objectives. Among the most important qualities are the following:

Interpersonal skills. Project managers must be able to relate positively to people. They must listen actively and be able to empathize with the needs of people.

Communicative ability. Project managers must be able to present their ideas both orally and in written form. Oral presentations occur frequently on projects and good presentation skills are absolutely necessary to motivate the team. Good writing skills are needed to prepare project documentation.

Decisiveness. Project managers must not avoid making tough decisions. Neither should they be impulsive in their judgments. They should, however, be willing to make timely decisions and accept responsibility for the consequences.

Objectivity. Project managers should be objective, especially when they receive important information that they would rather not hear. The opposite of objectivity is subjectivity, a quality that can quickly damage morale and project performance.

Commitment. Project managers should be totally committed to the success of the project. Commitment means providing the necessary technical, operational, and economic support to accomplish goals and objectives. Lack of commitment can become contagious to other participants, causing productivity can decline.

Motivation. Project managers should be able to get people involved in the project and sustain involvement until the goals and objectives have been achieved. If project managers can't motivate, then they can't captivate and, consequently, the team won't perform at a satisfactory level.

Honesty. Project managers must be forward, up-front, and truthful. If project managers fail in this regard, they will find managing their project very difficult. Credibility will inevitably decline, jeopardizing their authority.

Consistency. Project managers cannot deviate from the vision or the path unless extraordinary circumstances force them to. Project managers must make decisions that will help achieve project goals and objectives. Consistency breeds

10000 PROJECT LEADERSHIP

P² M² CYCLE — TASKS · RESPONSIBILITIES · INPUTS · OUTPUTS · MEASURES OF SUCCESS

TASKS

- 10100 Vision
- 10200 Communication
- 10300 Motivation
- 10400 Direction
- 10500 Support
- 10600 Team Building

MEASURES OF SUCCESS

- Does a vision exist?
- Is there a communications plan?
- Are incentives being used to motivate project participants?
- Is the project proceeding according to plan?
- Does the environment facilitate performance?
- Are techniques for effective team-building being used?

INPUTS (Tasks × Inputs)

TASK	Budget	Legal Considerations	Management Direction	Market Considerations	Mission Statement	Major Participants	Schedule	Statement of Work	Work Breakdown Structure	Monitoring	Status Assessment	Status Collection	Tracking	Management Style	Vision	Forms	Meetings	Newsletter	Procedures	Reports	Delegation	Incentives	Job Structuring	Participation
10100 Vision	X	X	X	X	X	X	X	X	X															
10200 Communication										X	X	X	X		X									
10300 Motivation														X	X									
10400 Direction															X	X	X	X	X	X				
10500 Support											X	X	X		X					X				
10600 Team Building															X						X	X	X	X

RESPONSIBILITIES (Tasks × Responsibilities)

TASK	Project Manager	Senior Management	Project Sponsor	Project Team	Client
10100 Vision	X	X	X	X	X
10200 Communication	X			X	X
10300 Motivation	X			X	
10400 Direction	X		X	X	X
10500 Support	X			X	
10600 Team Building	X			X	X

OUTPUTS (Tasks / Measures × Outputs)

TASK / MEASURE	Esprit de Corps	Productive Environment	On target	Motivated participants	Effective Communications	Focused Effort
10100 Vision						X
10200 Communication					X	
10300 Motivation				X		
10400 Direction			X			
10500 Support		X				
10600 Team Building	X					
Does a vision exist?						X
Is there a communications plan?					X	
Are incentives being used to motivate project participants?				X		
Is the project proceeding according to plan?			X			X
Does the environment facilitate performance?		X				
Are techniques for effective team-building being used?		X			X	

stability and enables participants to adapt to changing circumstances. Lack of consistency only breeds dissension.

Vision. Project managers must be able to see the end result, even if it isn't clear in everyone else's mind. They must have the ability to visualize where the project is going and make sure everything happens to achieve the project vision.

Proactive approach. Project managers do not wait around for their projects to "happen." They must take the initiative to keep the project moving forward according to plan. They must accept complexity and change and recognize that the key is to manage change, not react to it.

4.1   VISION (10100)

Project managers and the people they manage require some reason for performing their work. The reason is contained in the **vision**, which is a what the project hopes to achieve. The vision should include well-defined goals and objectives, and a path to achieve them.

Project managers have a number of tools and techniques for developing the vision. These include a statement of work, work breakdown structure, schedule, and budget. The statement of work identifies the goals and objectives. The work breakdown structure, schedule, and budget provide the path for accomplishing these goals and objectives.

10100 VISION

P² M² CYCLE — TASKS · RESPONSIBILITIES · INPUTS · OUTPUTS · MEASURES OF SUCCESS

TASKS	WBS	Statement of Work	Schedule	Major Participants	Mission Statement	Market Considerations	Management Direction	Legal Considerations	Budget	Project Manager	Senior Management	Project Sponsor	Project Team	Client	Vision (Output)
10105 Review the assessment of the project environment			■	■	■	■		■	■	■					■
10110 Consult with Project Sponsor		■	■	■	■	■			■	■		■			■
10115 Prepare the project vision	■	■		■	■	■	■			■					■
10120 Consult with major project participants	■	■	■	■	■					■	■				■
10125 Distribute the vision				■			■			■	■	■	■	■	■

MEASURES OF SUCCESS

	Vision (Output)
Is there a vision in place?	■
Is the vision being communicated?	■
Is the vision being followed?	■

4.2 COMMUNICATION (10200)

A vision is of little value if the project team is not aware of it. It is important to communicate the vision to everyone. This key task is often overlooked by project managers.

Often, project managers know and understand the project vision but treat the project team as "worker bees," who concentrate on tasks but have no idea how their output contributes to the accomplishment of project goals and objectives. Through **communication** comes greater awareness of the importance of one's work to the project's success.

But communicating the vision is not enough. Project managers need to communicate with team members about a wide range of issues. Whatever the topic under discussion, it is important to have the mechanisms in place to sustain communications throughout the course of the project.

Project managers have a wide range of mechanisms for establishing and maintaining communications. These include reports, forms, newsletters, meetings, and procedures. The idea is to communicate openly and honestly on a regular basis.

10200 COMMUNICATION

P² M² CYCLE (INPUTS, TASKS, RESPONSIBILITIES, OUTPUTS, MEASURES OF SUCCESS)

	INPUTS					TASKS	RESPONSIBILITIES					OUTPUTS					
	Vision	Tracking	Status Collection	Status Assessment	Monitoring		Project Manager	Senior Management	Project Sponsor	Project Team	Client	Reports	Procedures	Newsletters	Meetings	Forms	Communications Plan
TASKS																	
10205 Develop a project communications plan	■	■	■	■	■		■					■	■	■	■	■	■
10210 Document the plan	■						■										■
10215 Publish the plan	■						■			■	■						■
MEASURES OF SUCCESS																	
Does a communications plan exist?												■	■	■	■	■	■
Has the communications plan been implemented?																	■
Are communications tools in place?												■	■	■	■	■	

4.3 MOTIVATION (10300)

Motivation is absolutely essential to leadership. Without motivation on the part of the project manager, the team's ability to progress is severely hampered. Motivating involves encouraging people to participate actively to attain project goals and objectives.

A number of theories exist on motivating subordinates. These theories observe that people are motivated to fulfill physiological and psychological needs and desires. These needs and desires can be fulfilled both intrinsically or extrinsically, depending on the individual.

The project manager has several tools and techniques for motivating project participants. These include incentives, job structuring, participation, and delegation.

Positive or negative incentives can be used to motivate. Unfortunately, most project managers have little formal control over their resources. They often lack the formal power to provide financial incentives. However, they can manage the project in ways that will encourage people to perform satisfactorily or even exceptionally in achieving project goals and objectives.

Project managers can also restructure involvement in projects. They can, for example, assign tasks that will enrich people. They can also assign tasks to "stretch" or challenge people without demoralizing them. Doing both requires judgment by the project manager.

Project managers can encourage greater team participation in the affairs of the project. This participation can range from providing information and reviews to making decisions that influence the outcome of the project in terms of schedule, budget, and quality.

Finally, project managers can delegate tasks. Delegating not only increases greater participation in the project but also gives people an opportunity for growth.

When delegating, project managers must look at a number of factors, including the nature of the task being delegated, the attributes of the person to do the tasks, and the constraints on performing the work.

10300 MOTIVATION

INPUTS		P² M² CYCLE	RESPONSIBILITIES					OUTPUTS			
Vision	Management Style		Project Manager	Senior Management	Project Sponsor	Project Team	Client	Participation	Job Structuring	Incentives	Delegation
		TASKS									
▨		**10305** Determine the incentive at your disposal	▨		▨			▨	▨	▨	▨
▨		**10310** Develop ways to increase job satisfaction	▨		▨			▨	▨	▨	
▨	▨	**10315** Determine ways to encourage participation	▨		▨			▨	▨		▨
▨	▨	**10320** Determine the circumstances when to delegate	▨		▨						▨
		MEASURES OF SUCCESS									
		Do incentives exist to improve performance?								▨	
		Are measures in place to maintain or improve job satisfaction?						▨	▨		▨
		Have the groundrules been established and communicated for project team participation?						▨	▨		

4.4 DIRECTION (10400)

Project managers must maintain a consistent **direction** in the pursuit of project goals and objectives. Many project managers fall into the trap of forgetting about the overall goals and objectives of the project and start focusing on the technical details of the project. The results are myopia and straying from the path leading to those goals and objectives. The idea is to focus on the overall picture to ensure movement in the right direction.

Project managers have a number of tools and techniques for maintaining direction. These are monitoring, tracking, status collection, and status assessment. The idea is to employ these tools and techniques to help them see not only that they are staying on the path but also how well they are progressing.

10400 DIRECTION

INPUTS						P² M² CYCLE	RESPONSIBILITIES					OUTPUTS	
Vision	Reports	Procedures	Newsletter	Meetings	Forms		Project Manager	Senior Management	Project Sponsor	Project Team	Client	Corrective Action	Direction
TASKS													
■						10405 Focus on the vision	■		■	■			■
■	■	■		■	■	10410 Identify vision variances	■						■
■	■	■			■	10415 Determine if variances require corrective action	■					■	
■						10420 Take corrective action	■		■			■	
MEASURES OF SUCCESS													
						Have the goals and objectives been communicated?							■
						Are the goals and objectives reflected in project documentation?							■
						Do measures exist to determine whether your objectives are being accomplished?						■	■
						Are plans established to handle deviations to goals and objectives?						■	■

4.5 SUPPORT (10500)

Project managers facilitate, even expedite, the performance of their team. They provide value-added **support** to team members to assist in achieving the project goals and objectives. As leaders, project managers must do whatever is necessary to help project participants perform their tasks efficiently and effectively. They ensure that the right tools are available, that the work environment is physically free from distractions, and that administrative obstacles do not hinder work performance.

10500 SUPPORT

INPUTS					P² M² CYCLE	RESPONSIBILITIES					OUTPUTS	
Vision	Tracking	Status Collection	Status Assessment	Reports		Project Manager	Senior Management	Project Sponsor	Project Team	Client	Surmounted Obstacles	Right Tools
					TASKS							
✓	✓	✓	✓	✓	**10505** Determine the necessary tools to perform the project	✓			✓			✓
✓	✓	✓	✓	✓	**10510** Identify the obstacles in the work environment	✓			✓		✓	
			✓	✓	**10515** Identify areas that will involve red tape	✓					✓	
✓	✓	✓	✓	✓	**10520** Determine the overall needs to complete the project	✓			✓		✓	✓
					MEASURES OF SUCCESS							
					Have the major obstacles to performance been identified?						✓	
					Do plans exist for alleviating the impact of those obstacles?						✓	
					Are the right tools selected?							✓

The P² M² CYCLE diagram shows: INPUTS → TASKS → RESPONSIBILITIES → OUTPUTS → MEASURES OF SUCCESS (cyclical).

4.6 TEAM BUILDING (10600)

Team building in many ways is a reflection of the responsibilities and qualities of project managers. Yet, without an effective team, the likelihood of a successful project decreases.

A project manager with good team-building qualities typically performs several actions, including:

Appointing leaders
Assigning responsibilities
Encouraging camaraderie
Engendering enthusiasm
Instituting effective span of control
Instituting unity of command
Maintaining accountability
Providing a good work environment
Providing effective communications

10600 TEAM BUILDING

P² M² CYCLE

TASKS	Project Manager	Senior Management	Project Sponsor	Project Team	Client	Span of Control	Authority	Responsibility	Accountability	Esprit de Corps	Team	Delegation	Incentives	Job Structuring	Participation	Vision
10605 Determine the overall structure of the team	X			X		X	X	X	X		X				X	X
10610 Establish a reporting structure	X			X			X	X			X			X	X	
10615 Ensure the team structure has effective span of control	X			X		X	X				X			X		
10620 Ensure the team structure has no one working for more than one lead	X			X		X			X		X			X		
10625 Identify ways to build cameraderie	X									X	X	X	X	X		
10630 Ensure good accountability in assignments	X			X		X	X	X	X		X			X	X	
10635 Appoint team leads	X					X	X	X	X		X			X	X	
MEASURES OF SUCCESS																
Does an organizational structure exist?					X											
Has the organizational structure been communicated?										X	X					
Does everyone know their responsibilities?											X					
Does the organizational structure encourage effective span of control?						X					X					
Does the existing structure follow the unity of command principle?										X	X					

(Columns grouped as: **RESPONSIBILITIES** — Project Manager, Senior Management, Project Sponsor, Project Team, Client; **OUTPUTS** — Span of Control, Authority, Responsibility, Accountability, Esprit de Corps, Team; **INPUTS** — Delegation, Incentives, Job Structuring, Participation, Vision)

5

Project Definition (20000)

Introduction

Direction is vital for successfully completing projects on schedule, within budget, and of a high quality. The best way to start a project in the right direction is to define the goals and objectives early.

Goals are meaningful statements describing what the project will achieve. Goals are not generally measurable.

Objectives are measurable subsets of goals. The attainment of objectives gives a good indication of how well the project achieves the overall goals.

The project manager should formalize the goals and objectives by incorporating them into a statement of Work (SOW), or project charter. An approved SOW reflects the project manager's, client's, and project sponsor's commitment and acceptance of the goals and objectives. The SOW also formalizes other project parameters, such as ones related to cost, schedule, and quality.

An important aspect of defining the project is to define the roles and responsibilities of the participants. This task will be reviewed and updated throughout the project as new phases are started and new resources are utilized.

A project announcement should also be made. This document communicates what the project hopes to accomplish and who the major participants are. Immediately after the project announcement, a project launch session should be held to bring the team members together to officially start the project.

Project definition is absolutely critical to the success of a project. Failure to define the project satisfactorily can result in poor performance, waste, and ineffectiveness.

20000 PROJECT DEFINITION

P² M² CYCLE (TASKS — RESPONSIBILITIES — OUTPUTS — MEASURES OF SUCCESS — INPUTS)

TASKS

- 20100 Goals and objectives determination
- 20200 Statement of Work
- 20300 Project announcement
- 20400 Project launch
- 20500 Roles and responsibilities

MEASURES OF SUCCESS

- Are there goals and objectives for the project?
- Is there a Statement of Work?
- Has a project announcement been published?
- Have the roles and responsibilities for project participants been documented? Have they been communicated to participants? Have they been agreed to?
- Has the project launch meeting been held?
- Is the Statement of Work signed off?

TASK	INPUTS: SoW	Major Participants	Assumptions	Constraints	Major Tasks	Major Deliverables	Major Responsibilities	Project Objectives	Project Goals	Mission Statement	Market Considerations	Management Direction	Legal Considerations	RESP: Project Manager	Senior Management	Project Sponsor	Project Team	Client	OUT: Distribution	Project Sponsor Signature	Project Sponsor	Project Announcement	Procedures	Measurement Criteria	Statement of Work	Project Objectives	Project Goals
20100 Goals and objectives determination		■	■	■	■	■	■	■	■					■	■	■	■	■					■		■	■	■
20200 Statement of Work	■									■	■	■	■	■	■	■	■	■				■		■	■	■	■
20300 Project announcement		■					■							■		■	■	■	■	■	■	■					
20400 Project launch		■	■		■	■	■	■	■	■				■		■	■	■	■	■	■	■			■	■	■
20500 Roles and responsibilities							■							■		■		■									
Are there goals and objectives for the project?																										■	■
Is there a Statement of Work?																									■		
Has a project announcement been published?																				■	■	■	■			■	■
Have the roles and responsibilities for project participants been documented? Have they been communicated to participants? Have they been agreed to?																			■	■							
Has the project launch meeting been held?																									■	■	■
Is the Statement of Work signed off?																									■		

5.1   GOALS AND OBJECTIVES DETERMINATION (20100)

Common sense tells us projects should all have a purpose for their existence. Ironically, this basic premise is often overlooked. Many projects have no defined goals or objectives. The project manager or team members—or both—do not have the slightest idea of the purpose of their project. Instead, they come to work and simply do whatever they must to earn their paycheck.

Projects with No Direction

This scenario is more common than many project managers will admit. The consequences are disastrous; they are exemplars of a reactive style of project management.

Projects that proceed without any direction have the pattern of a bag of marbles falling to the floor: every action resembles marbles scattering in different directions. Energy, consequently, is dissipated. People perform tasks in nonproductive ways. Duplicating effort and repeating tasks are common manifestations of a project plagued with unclear or no goals and objectives.

Managers of such projects poorly employ resources, such as people, equipment, supplies, or facilities. Resources used in this environment may be misallocated. For example, personnel who have no need for personal computers may be issued them anyway; meanwhile, other people, who need the personal computers, go without them. Consequently, productivity declines as project costs increase.

Projects without goals and objectives lead to poor morale. People need a purpose for what they are doing; goals and objectives—definable ones—are the first paths towards giving people purpose. If people feel like they are "spinning their wheels," they become restless and go someplace else for direction. Goals and objectives give people meaning to their work; they give people a mission.

Finally, projects without goals and objectives lead to dissension. People involved with the project—client, project sponsor, or project team—will likely start disagreeing among one another to the point that the project accomplishes nothing. Projects in this atmosphere usually lead to an impasse, making progress difficult, if not impossible.

Value of Having Goals and Objectives

In addition to recognizing the value of goals and objectives, it is necessary to clearly and precisely define them.

A goal is a statement of intent. It clearly indicates the direction in which an organization is going. A goal is definable but is not measurable by itself.

An objective is a subset of a goal; it is measurable and supports the attainment of one or more goals. Objectives are benchmarks telling you whether you are reaching your goals. **Exhibit 5.1-1** shows the relationship between goals and objectives.

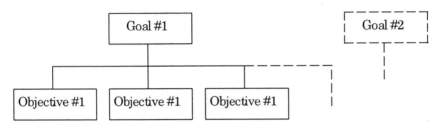

RELATIONSHIP BETWEEN GOALS AND OBJECTIVES
EXHIBIT 5.1-1

The following is an example of a goal:

Build a small, state-of-the-art twin-engine jet aircraft within schedule and budget.

The following are objectives supporting the goal:

The craft will carry up to 40 passengers.
The craft will weigh 20% less than current crafts of similar make and size.
The craft can travel up to 1500 knots.
The craft will consume 37% less fuel than current crafts of similar make and size.
The project will finish under the budget allocation of $20 million.
The project will finish by August 19, xxxx.

Notice the goal is specific enough to tell you what you are seeking to achieve but is not measurable. Yet, you know what direction you are taking: to build a twin-engine jet.

The objectives give something measurable. These are benchmarks supporting the achievement of the goal. They are, quite literally, yardsticks telling you how well you must perform.

You can have multiple goals and objectives for a project. However, every project must have one overriding goal; that is, what is it that you are going to achieve? You can break the highest level goal into "smaller" goals, which in turn, can be broken into smaller ones. Likewise, you can do the same with objectives;

the only difference is that you do not have one overriding objective. In turn, each goal can have one or more objectives, while each objective can support the attainment of one or more goals.

Remember that your objectives must support the attainment of goals. In other words, the latter must support the attainment of the former. You should avoid having objectives that do not support accomplishing a goal(s) established for the project.

20100 GOALS AND OBJECTIVES DETERMINATION

	INPUTS			P² M² CYCLE / TASKS	RESPONSIBILITIES					OUTPUTS				
Mission Statement	Market Considerations	Management Direction	Legal Considerations		Project Manager	Senior Management	Project Sponsor	Project Team	Client	Measurement Criteria	Statement of Work	Project Objectives	Project Goals	Project Charter
TASKS														
■	■	■	■	20105 Obtain higher level direction from various sources	■	■	■					■	■	
■				20110 Determine when to perform goals and objectives definition	■							■	■	
■				20115 Determine where to perform goals and objectives definition	■							■	■	
■				20120 Assign responsibilities for performing goals and objectives definition	■		■					■	■	
■	■	■	■	20125 Reflect statement of goals and objectives in project charter and SOW	■						■			■
■	■	■	■	20130 Determine goals of project	■	■	■		■		■		■	
■	■	■	■	20135 Determine objectives of project	■	■	■		■			■		
■		■		20140 Determine measurable criteria for attainment of objectives	■					■				
■				20145 Communicate goals to project team	■		■						■	
■				20150 Communicate objectives to project team	■		■					■		
MEASURES OF SUCCESS														
				Have the goals and objectives been defined adequately?							■	■	■	
				Have the goals and objectives been communicated to all project participants?							■	■	■	■
				Are the objectives measurable?						■		■		
				Are the goals and objectives "in synch" with higher organizational goals and objectives?								■	■	
				Is there a measurement criteria available to determine achieving goals and objectives?						■				

5.2 STATEMENT OF WORK (20200)

Sometimes, disagreement will occur over what the overriding goal of the project is. The best tool to resolve this problem is the **Statement of Work (SOW)**.

The SOW is a document defining the scope of the project and the responsibilities of the participants. It offers several advantages.

First, it serves as a mechanism to identify points of agreement among project participants. When drafting the SOW, the customer, project sponsor, and project manager can formally agree on the goals and objectives of the project. They can also agree on who is responsible for doing what.

But the SOW serves another, often overlooked, purpose: it forces disagreements to surface early. These very disagreements might otherwise be ignored until later in the project. The process of drafting the SOW forces issues into the open and brings early resolution to them.

The SOW serves as a formal contract between the project manager, project sponsor, and customer. When all parties sign it, they're agreeing to comply with its contents. It can even be stipulated that noncompliance will result in direct performance penalties.

Finally, the SOW defines the scope of the project. It defines the product for delivery to the customer. Defining the scope involves documenting project constraints (i.e., mandated completion date, limited budgets, and resources available).

Statement of Work Components

The following are the primary components of the SOW:

Introduction. This section of the SOW briefly describes the project. It is a description explaining the purpose of the project and the major participants. It may include some historical information regarding what led to the project.

Goals and objectives. The goals and objectives identified earlier need to be included in the SOW. A reader reviewing this section should clearly understand the purpose of the project, what specifically will be delivered, and how success will be measured (through the objectives).

Scope. The scope defines the boundaries of the project. The scope should be defined in terms of what is included in the project and what is excluded.

Critical planning assumptions. In preparing the SOW, the project manager will have made several key assumptions. These assumptions should be documented and agreed upon by the project sponsor and the project manager. As the project progresses, some of the these assumptions may prove to be in error and may require some replanning of the project.

Key stakeholders. Key stakeholders of the project need to be identified. They will represent individuals or groups whose interests need to be considered throughout the project. They may be internal or external resources that need to be part of the project manager's communications plan.

Project resources. This section should identify the key resources involved in the project along with the responsibilities that each of these resources (or groups) will undertake. In cross-functional projects, it is critical that all areas be identified in this section. It is a mistake for project managers to include just the resources within their own department.

Milestone schedule. A schedule of project milestones should be presented as part of the SOW.

Budget. The budget for the project should be included in the SOW. If possible, the budget should be broken down by major milestones to show some degree of cash flow timing for the project.

Amendments. This section provides the steps all parties agree to follow when seeking a variance to the SOW. It provides an orderly mechanism for incorporating those changes.

Signatures. This final section records the date and signatures of the customer, the project sponsor, and the project manager. The signatures and date serve as proof that all parties agree to comply with the contents of the SOW.

Exhibit 5.2-1 is an example of an SOW following this structure. Knowing the structure and contents, however, is just one facet of developing a SOW. You also have to arrive at an agreement between yourself and the customer; you may find that reaching consensus is much harder than writing the SOW.

Statement of Work

Project: Move Marketing Department
Project Manager: Bill Smith
Project Sponsor: Jane Young - Vice President, Marketing

Introduction
The ABC currently has staff distributed throughout the city. The plan is to consolidate all corporate departments into the new corporate headquarters.

Goals and Objectives
The goal of the project is to move the Marketing Department from its current location at 700 Main Street to the new corporate headquarters at 800 Georgia Street. The project is to be completed within the $200,000 budget, with the move completed by March 15th, and to have no more than 2 days downtime for staff as a result of the move.

continues

Scope
The scope is to plan for the move, select appropriate vendors to execute the move, and manage the actual move. The move includes only the Marketing Department; other corporate departments at 700 Main Street are not included in this project.

Critical Planning Assumptions
The team has assumed the new building will be ready by March 1st in order to facilitate the move.

Key Stakeholders
The marketing staff, customers, and other parts of the organization that rely on marketing staff input are the key stakeholders.

Project Resources
The project will consist of a cross-functional team of six internal staff and several vendors who will be selected as part of the project. Internal resources will come from Marketing, Information Technology, Mail Room and Human Resources. The external staff will come from the selected moving company.

Milestone Schedule
The following are the key milestones for the project:
- Initiate project   November 1
- Document requirements   November 30
- Select mover   January 15
- Complete move plan   February 15
- Move to new building   March 15

Budget
The budget for the project is $200,000.

Amendments
All changes to this statement of work will be processed through a change management process. Approvals will be required prior to the making any changes which affect this document.

Approvals

Project Sponsor	Project Manager	Customer
Date	Date	Date

STATEMENT OF WORK EXAMPLE
EXHIBIT 5.2-1

Steps for Building an SOW

Getting to the point of agreeing to an SOW is a difficult and time-consuming task. Ironically, it requires only a few steps, but actually completing them is very difficult. **Exhibit 5.2-2** illustrates the steps.

ACTIONS FOR DEVELOPING A STATEMENT OF WORK
EXHIBIT 5.2-2

The first step is for the project manager to draft the Statement of Work. As a project manager, you should write what you think should go into the document. Remember, the project manager is the one who takes the initiative in starting a project in the right direction.

Submit a copy for review to the customer(s) and project sponsor. Give them ample time to review the contents and debate over the contents. They will likely have revisions.

Hold a meeting to discuss discrepancies and resolve differences over the SOW. Be sure to document the meeting and distribute these minutes to all attendees. The goal of the exercise is not to "cover yourself" but to ensure that all participants are aware of the decisions which have been agreed to in the meeting.

These efforts will help you to develop an SOW which is realistic, reasonable, and acceptable to all.

Revise the SOW by making changes around the issues discussed in the meeting. Do not unilaterally make changes, otherwise you could find yourself in constant battle. Seek consensus as early as possible.

Hold another review session to discuss any remaining problems with the SOW. If no problems exist, all parties should date and sign the SOW. If problems exist, reschedule the meeting and make the necessary revisions that will lead to consensus. If you find disagreement again, repeat scheduling a review session and make the revisions until all parties agree to sign the SOW.

You should **never** begin a medium or large project without a completed and approved Statement of Work. Proceeding without a signed SOW indicates that the customer does not have a clear idea of what it wants, and this will dramatically affect your ability to manage the project.

Beginning a project without an SOW will lead you to operate based upon some assumptions. You might expect that the customer wants one thing, only later to find (after expending considerable time and effort) that it wants something else. Under such circumstances, you and your team will perform less productively and waste resources. In the end, you may have to absorb the total cost of the effort.

Pitfalls to Avoid When Developing an SOW

Sometimes project managers draft and agree to less-than-satisfactory Statements of Work. These documents contain serious shortcomings that can later lead to political and legal complications.

Project managers will sometimes cave in to an unrealistic request made by the customer and recorded in the SOW. They may have agreed to a provision to keep the customer satisfied to ensure the project continues. Later, they may fail to comply with the provision, which can lead to embarrassing complications.

Including phrases that lack clarity is another shortcoming of some Statements of Work. These phrases can quickly lead to misinterpretation and make it impossible for the project to proceed smoothly.

It does not help if the Statement of Work is signed by all parties, only to be placed in a desk drawer and never reviewed. During the project, deviations to the provisions may occur; either party may be unaware of such occurrences. This unawareness, however, should not be construed as concurrence by the customer.

Finally, Statements of Work are sometimes incomplete. They may fail to delineate all the constraints facing the project. These constraints may relate to the budget, schedule, and quality. They may also include any legal, regulatory, policy, or procedural constraints placed on the project. Useful Statements of Work attempt to list **all** constraints placed on a project.

When You Need an SOW

Not every project requires an SOW. Small, internal projects may not require one. If you are both the customer and the project manager (e.g., you request and perform an audit), you do not need an SOW. Under certain circumstances, you also do not need to develop an SOW for working on projects with your immediate boss unless the latter requests it; these projects may be small enough that you alone are the project team.

Whether or not you need a Statement of Work depends on the size and complexity of the project. You should develop a Statement of Work whenever the project is of a magnitude that it requires a separate project manager, customer, and project team.

20200 STATEMENT OF WORK

Major Participants	Assumptions	Constraints	Major Tasks	Major Deliverables	Major Responsibilities	Project Objectives	Project Goals	P² M² CYCLE / TASKS	Project Manager	Senior Management	Project Sponsor	Project Team	Client	Procedures	Statement of Work
			■	■	■	■	■	20205 Draft outline of SOW	■						■
	■	■	■	■	■	■	■	20210 Determine length	■						■
						■	■	20215 Determine scope	■	■	■	■			■
					■			20220 List responsibilities	■		■	■			■
■								20225 Determine who approves	■	■	■	■		■	■
■	■	■	■	■	■	■	■	20230 Determine contents and kinds of amendments	■		■	■			■
			■	■		■	■	20235 Determine what goes into the introduction	■						■
		■						20240 Include all key constraints	■	■	■		■		■
■								20245 Assign responsibilities for preparing the SOW	■					■	■
						■		20250 Institute version control	■					■	■
■						■		20255 Establish a revision process	■					■	■
■	■					■		20260 Determine the number of review sessions required	■						■
	■	■						20265 Determine the pressures to accept an unrealistic SOW and the ways to deal with those pressures	■						■
■								20270 Determine distribution	■	■	■	■	■		■
								MEASURES OF SUCCESS							
								Is the SOW complete?							■
								Is the SOW clear?							■
								Have all the principal participants approved the SOW?							■
								Does the SOW contain the goals and objectives?							■
								Have all participants seen the SOW?							■
								Are changes to the SOW managed?						■	■

5.3 PROJECT ANNOUNCEMENT (20300)

A **project announcement** is a brief, one-page memo announcing the existence of a project. It also announces the goals of the project and its project manager. The person signing the project announcement is the project sponsor. The sponsor is the person who will support the project financially and politically.

The project announcement offers several advantages. One, it communicates to everyone that a new project is underway and who the project manager is. The project announcement also reveals the project sponsor. This announcement becomes politically important when different projects compete for resources.

Not surprisingly, project sponsors will struggle over the wording in the project announcement. The project announcement should be signed by the project sponsor and distributed to the key stakeholders and other affected parties. The letter should include:

> Project name
> Project goal
> Project manager
> Major participants
> Request for support
> Sponsor's signature

Exhibit 5.3-1 shows an example of a project announcement letter:

To: Finance Staff   June 11, xxxx
     Customer Distribution Staff
cc: Vice President, Finance

SUBJECT: AMIS Project

Effective immediately, a project team has been formed to develop an Asset Management Inventory System (AMIS) for the Customer Distribution Division. Cheryl Smith-Jones is the project manager. Cheryl will work closely with the Division's Manager, Bill Glassersen.

I look forward to your participation in the success of this project.

Janice Widmore
Vice President, Customer Distribution

SAMPLE PROJECT ANNOUNCEMENT LETTER
EXHIBIT 5.3-1

20300 PROJECT ANNOUNCEMENT

P² M² CYCLE

	INPUTS		RESPONSIBILITIES					OUTPUTS			
TASKS	Major Participants	Statement of Work	Project Manager	Senior Management	Project Sponsor	Project Team	Client	Distribution	Sponsor Signature	Project Sponsor	Project Announcement
20305 Identify Project Sponsor		▦	▦	▦						▦	
20310 Draft the announcement		▦	▦								▦
20315 Review with Project Sponsor and obtain Project Sponsor's signature		▦	▦		▦				▦		
20320 Distribute Project Announcement	▦		▦	▦	▦		▦	▦			
MEASURES OF SUCCESS											
Has a Project Sponsor been identified?										▦	
Is that person signing the announcement?									▦		
Has the announcement been distributed to the right people?								▦			
Does a project announcement exist?											▦

5.4 PROJECT LAUNCH (20400)

The **project launch session** is the formal recognition of the start of the project. The project launch session should be attended by all team members, key stakeholders and key committee members. The project manager should review the statement of work, discuss the importance of working as a team, and establish an open and friendly attitude within the team. The project sponsor should address the group about the business benefits of the project, create a sense of excitement and enthusiasm within the team, and show personal support for the project.

The launch session is an opportunity for all team participants to introduce themselves, get to know the other people they will be working with on the project, and generally to start the team-bonding process. The launch session should help people relax and facilitate this critical team-bonding experience. Obviously, it should be held at the beginning of the project. In addition to this session, the project manager should consider the potential to have subsequent "launch" sessions, perhaps for the beginning of a new phase of the project or when a significant number of new people are joining the team.

20400 PROJECT LAUNCH

	INPUTS		P² M² CYCLE		RESPONSIBILITIES					OUTPUTS		
Major Participants	Statement of Work			Project Manager	Senior Management	Project Sponsor	Project Team	Client	Kickoff Meeting	Team Bonding	Communicated Shared Vision
			TASKS								
		■	**20405** Determine who to invite	■		■				■	
■	■	■	**20410** Formally invite participants	■						■	
■			**20415** Have Project Sponsor attend	■		■					■
	■	■	**20420** Hold kickoff meeting	■		■	■	■	■	■	■
			MEASURES OF SUCCESS								
			Has a Project Sponsor provided a motivational talk to participants?							■	■
			Has the project team started bonding?							■	
			Are the team members getting to know each other?							■	
			Has there been a meeting to launch the project?						■		

5.5 ROLES AND RESPONSIBILITIES (20500)

A critical activity for the project manager and project sponsor is to identify who will be involved in the project and to clearly document and communicate the **roles and responsibilities** of these individuals.

Chapter 3 introduced the key participants in a project. The following are the key roles and responsibilities for a project. The project manager will have to define the key project participants for his or her own specific projects and should work with these key resources to develop appropriate roles and responsibilities for each of them.

Project Sponsor

The **project sponsor** is assigned to the project by senior management. The sponsor is ultimately responsible for the overall success of the project. The key roles and responsibilities are:

- Assigning the project manager
- Establishing the business objectives of the project and ensuring that these objectives are met
- Acquiring sufficient resources to ensure the success of the project
- Signing legal contracts, as required
- Reviewing and resolving funding requirements that are outside the project commitment
- Reviewing and resolving decision and change requests
- Authorizing all changes to the statement of work
- Signing off on key project deliverables
- Executing final sign-off of the project

Project Manager

The **project manager** has direct responsibility for managing the delivery of the project as identified in the Statement of Work. The key roles and responsibilities are:

- Successfully completing the project
- Understanding the customer requirements
- Understanding and managing the project within the scope identified in the SOW
- Managing the project to accomplish the goals and objectives identified in the SOW
- Providing status reports to the project sponsor and other key stakeholders
- Identifying and acquiring resources necessary to complete the project

- Ensuring the quality and content of all project deliverables
- Using change management practices to manage all changes to the project
- Managing and controlling the project plan, resources, quality, and costs

Project Team

The **project team** consists of the individuals who complete tasks and produce deliverables for the project. The key roles and responsibilities of the project team members are:

- Providing input to the planning process in terms of tasks required, deliverables, and estimates
- Completing tasks as identified in the project plan
- Reporting status to the project manager
- Identifying changes or decisions as early as possible

Client

The **client** represents the person(s) or organization(s) that will be the recipients of the project deliverables. The key roles and responsibilities of the client are:

- Providing input on the client requirements
- Providing the team enough information to ensure success
- Reviewing all deliverables produced by the team
- Participating in acceptance testing of deliverables, where appropriate
- Signing off on deliverables

Senior Management

Senior management is responsible for determining which projects will be initiated. On specific projects, the key roles and responsibilities are:

- Assigning the project sponsor
- Reviewing and resolving any project related issues that are directed to senior management
- Considering the impact of the project on other corporate projects and activities

Client Review Committee

The **client review committee** is formed for some projects to review project deliverables and to provide client acceptance to the project. This committee is common on large projects or projects where the deliverables will be utilized across multiple organizations. The key roles and responsibilities are:

- Providing executive input on business requirements
- Reviewing of project deliverables

- Testing of project deliverables
- Signing off of project deliverables

Project Steering Committee

The **project steering committee** is formed in some projects to provide direction to the project team. This committee is usually formed on large projects or projects that affect multiple departments, divisions or organizations. The key roles and responsibilities are:

- Reviewing project status
- Ensuring project is within scope
- Providing guidance on issues related to risk management
- Reviewing and resolving appropriate project decision requests
- Reviewing and advising on project change requests

The chart in **Exhibit 5.5-1** shows the relationship among the key project participants. In addition to documenting the roles and responsibilities, the project manager should prepare a similar chart showing the relationships.

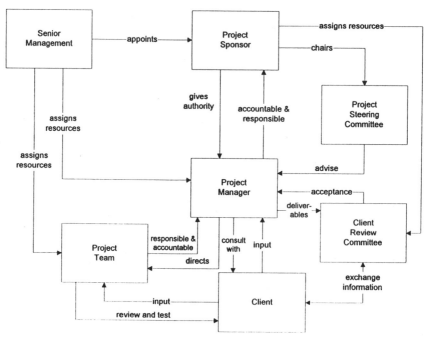

PROJECT RELATIONSHIPS
EXHIBIT 5.5-1

20500 ROLES AND RESPONSIBILITIES

INPUTS		P² M² CYCLE	RESPONSIBILITIES					OUTPUTS
Major Participants	Statement of Work		Project Manager	Senior Management	Project Sponsor	Project Team	Client	Roles and Responsibilities
		TASKS						
	▓	**20505** Identify key groups / functions to be involved	▓		▓			▓
	▓	**20510** Draft initial roles and responsibilities	▓					▓
	▓	**20515** Review with sponsor	▓		▓			▓
▓	▓	**20520** Review with key project participants	▓		▓	▓	▓	▓
▓	▓	**20525** Finalize roles and responsibilities	▓					▓
▓		**20530** Distribute to major project participants	▓					▓
		MEASURES OF SUCCESS						
		Have the roles and responsibilities been documented?						▓
		Have the roles and responsibilities been communicated?						▓
		Are the roles and responsibilities agreed upon?						▓

6

Project Planning (30000)

Introduction

After defining what the project will achieve, the next step is to determine how it will accomplish its goals and objectives. The way to accomplish goals and objectives is to create a work breakdown structure, develop time estimates, develop schedules, allocate resources, calculate costs, and manage risk.

Work breakdown structures (WBS). The WBS is a hierarchical listing of the products, subproducts, tasks, and subtasks required to complete the project. A good WBS provides the basis for developing meaningful time and cost estimates as well as useful schedules.

Time estimates. Project managers can apply time estimates against the tasks and subtasks identified in the WBS. There are a number of estimating techniques that can be applied depending on the degree of confidence they want in the estimates.

Schedules. Project managers can apply the outputs from the WBS and time estimates to develop schedules. They first identify logical relationships among tasks and apply time estimates against those tasks. Next, they calculate the dates for each task while keeping in mind the constraints placed upon the project. Through schedule calculation, project managers identify critical tasks for completing the project on time.

Resource allocation. Projects consume resources such as people, supplies, materials, equipment, and floor space. Project managers must assign resources to tasks to complete them. After applying resources, project managers can determine whether sufficient resources exist to complete the deliverables.

Cost calculation. After creating a work breakdown structure; determining time estimates; developing schedules; and allocating resources, project managers can calculate the cost to perform each task and for the entire project. The estimated cost eventually becomes the budget. While executing the project, the project manager tracks cost performance against the budget.

Risk control. No project occurs in a vacuum. A barrage of threats can influence how well a project proceeds and succeeds. An effective project manager attempts to determine those threats and develops a realistic plan to minimize their impact. A project manager, knowing the threats, can develop plans to minimize their occurrence or lessen their impact on the project.

30000 PROJECT PLANNING

P² M² CYCLE — TASKS · RESPONSIBILITIES · OUTPUTS · INPUTS · MEASURES OF SUCCESS

OUTPUTS

Output	30100 Work breakdown structure	30200 Estimating	30300 Risk control	30400 Schedules	30500 Resource allocation	30600 Cost calculation	Does a work breakdown structure exist?	Are reliable time and cost estimates available?	Has risk control been conducted?	Have resources been allocated?	Does a realistic schedule exist?
WBS	X						X				
Procedures	X	X	X	X		X	X	X	X		
Time Estimates		X						X			
Risk Identification			X						X		
Risk Assessment			X						X		
Risk Management			X						X		
Bar Chart				X							X
Network Diagram				X							X
Resource Histograms					X					X	
Revised Schedule					X						X
Resource Analysis					X					X	
Cost Estimates						X		X			
Management Reserve						X		X			

RESPONSIBILITIES

Responsibility	30100	30200	30300	30400	30500	30600
Client	X	X	X	X	X	X
Project Team	X	X	X	X	X	X
Project Sponsor		X		X	X	
Senior Management	X			X	X	
Project Manager	X	X	X	X	X	X

TASKS

- 30100 Work breakdown structure
- 30200 Estimating
- 30300 Risk control
- 30400 Schedules
- 30500 Resource allocation
- 30600 Cost calculation

MEASURES OF SUCCESS

- Does a work breakdown structure exist?
- Are reliable time and cost estimates available?
- Has risk control been conducted?
- Have resources been allocated?
- Does a realistic schedule exist?

INPUTS

Input	30100	30200	30300	30400	30500	30600
Major Participants	X	X	X	X	X	X
Statement of Work	X	X	X	X	X	X
Sources of Information	X					
Risk Identification		X		X	X	X
WBS		X		X	X	X
Political Pressures		X		X		
Management Direction			X			X
Potential Risks			X			
Allocated Resources					X	
Time Estimates				X	X	X
Bar Chart					X	
Network Diagram					X	
Resources					X	X
Revised Schedule						X
Cost Information						X

6.1 WORK BREAKDOWN STRUCTURE (30100)

Knowing your goals and objectives, while important, is not enough to complete your project. You must know what tasks are required to reach your goals and objectives. One mechanism for decomposing your goals into tasks is to use a **work breakdown structure** (WBS).

Definition of a WBS and Its Benefits

The WBS is a detailed listing of steps required to complete a project. It offers several benefits to project managers. Building a WBS forces project managers to think hard about what they must do to finish their projects. If done correctly, it leads to defining the exact steps to complete a project.

A WBS establishes solid groundwork to make realistic time and cost estimates for a project. Detailed listings of tasks facilitate calculating costs for individual activities and the entire project. By making time estimates for detailed tasks, project managers can "roll up" the data associated with each level of the WBS to build a composite picture of a project.

A good WBS also allows project managers to build accountability among project team members. After listing all tasks, project managers can assign people to each one. This helps to engender a sense of responsibility and accountability among team members.

Project managers can use the WBS to construct useful schedules. A definitive list of tasks paves the way for realistic estimates and the eventual construction of schedules. In addition, it enables project managers to develop meaningful summary schedules, known as tier schedules.

Finally, building a good WBS forces significant issues to arise early in a project rather than later (when it is more difficult to change the situation). To build the WBS requires considerable contributions from project participants (i.e., project manager, client, team members, project sponsor, and senior management).

Characteristics of a WBS

A major characteristic of the WBS is its top-down orientation. The project manager starts with the overall product and breaks it down into smaller elements. **Exhibit 6.1-1** shows this organization of a WBS.

The breakdown is very similar to preparing an outline for an essay. Each topic is broken down into subtopics, and each subtopic is further subdivided into components.

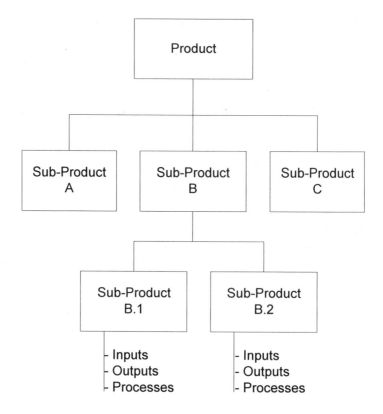

TYPICAL WORK BREAKDOWN STRUCTURE
EXHIBIT 6.1-1

Another characteristic is that the WBS is broken into several levels. Not all the "legs" of the WBS need explosion down to the same level. The level itself is of no major significance. Each level simply allows you to create schedules and reports summarizing information at each of those levels.

The WBS reveals **what**, not **how**; the sequence of each task is immaterial. Although people are used to reading from left to right, no sequence need be inferred. If you think sequentially, building the WBS can become very frustrating. It forces you to think about what needs to be done rather when it needs to be done. You can input the how and when for the tasks after constructing the WBS; that is, when you develop the schedule.

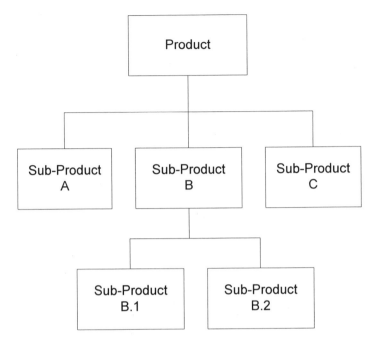

TYPICAL PRODUCT BREAKDOWN STRUCTURE
EXHIBIT 6.1-2

WBS Components

The WBS comprises two primary components. The **product breakdown structure** (PBS) is the first component of the WBS. The PBS, like the WBS in general, requires taking a top-down perspective. **Exhibit 6.1-2** illustrates the product breakdown structure.

Again, the breakdown of the PBS is similar to an outline for preparing an essay—each topic is broken into subtopics and each subtopic is further subdivided into components. The level of explosion depends on the complexity of the product. Generally, the more complex the product, the greater the number of levels.

The PBS also occupies the upper portion of the work breakdown structure. Each leg of the PBS is further subdivided into different levels. The overall product and each subproduct are described by a noun.

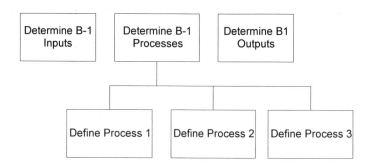

TYPICAL TASK BREAKDOWN STRUCTURE
Exhibit 6.1-3

The **task breakdown structure**, also known as the TBS, is the second component of the WBS. It comprises tasks, subtasks, and so on that are required to build each subproduct and ultimately to the building of the overall product.

The TBS, like the PBS, is broken down into several levels and requires a top-down perspective. **Exhibit 6.1-3** shows the task breakdown structure.

The level of explosion depends on the complexity of the overall product or subproduct. Generally, the more complex the project, the more levels are required.

The TBS, unlike the PBS, occupies the lower portion of the WBS. Each leg of the TBS can be subdivided down to different levels. Each task and lower element is described with a command verb (action verb) and an object. Hence, DEFINE (command verb) PROCESS 2 (object).

In certain companies, the PBS portion is constructed first and an active verb is added in front of the noun. Each task is subsequently exploded to more detailed levels. Another characteristic is that each item in the WBS (both the PBS and TBS portions) has a unique number, or code. The number identifies the position, or level, of the element within the WBS. Consider the example shown in **Exhibit 6.1-4**.

No element listed in the WBS has the same number or code, and as you proceed down to the lower levels, the numbers become larger. The numbering scheme reflects what element is a subset of a higher level element. For instance, DEFINE PROCESS 2 (2.1.2.2) is a subset of DETERMINE B-1 PROCESSES (2.1.2), which in turn is a subset of SUBPRODUCT B-1 (2.1) and so on. The lower an element is in a WBS, the longer the WBS number.

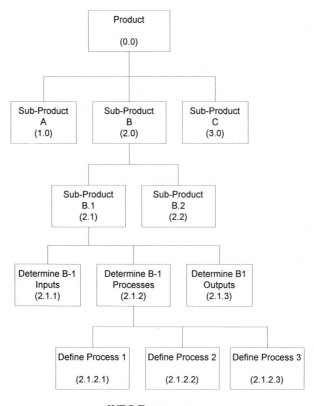

WBS EXAMPLE
EXHIBIT 6.1-4

Steps for Building a WBS

Keeping the aforementioned characteristics in mind, you can begin developing a WBS. Though only five steps are required, do not be fooled. Building a good, useful WBS takes hours—even days—of hard work and revisions.

Step 1. Write down the overall product you will build. Use a straightforward, descriptive noun or term, such as Inventory Management System or Marketing Plan. Be descriptive without being wordy. The source for the exact description of the product should come from the Statement of Work.

Step 2. Explode the overall product down to varying levels of subproducts. This helps build the product breakdown structure of the WBS. Do not worry if you break each leg down to a different level. You make the judgment how far you want to breakdown each leg. Usually, two or three levels will suffice.

Step 3. After you complete exploding the PBS portion, you can start doing likewise for the TBS portion by writing a series of next-level tasks under each lowest PBS element. The tasks within each given leg will be on the same level. Then, where applicable, explode each task down to lower levels. Again, each leg of the TBS can vary according to the number of levels.

A common question to ask is how far you should explode a leg on the WBS? The most common way to determine the appropriate level is known as the **2 week, or 80 hour, rule**. This means that if a TBS element requires more than two weeks of work, or 80 working hours, then explode the TBS element down another level. This rule of thumb will assure more detailed identification of tasks and will assist in tracking these tasks.

Step 4. Give each element within the WBS a unique numeric code. First, give the overall product element a code of 0.0 (a common practice). Second, at the next level of the PBS, give each subproduct element a unique whole number, typically 1.0, 2.0, etc. Using these numbers and starting downward on each leg, write the unique WBS code. The code should indicate its level of subordination to a higher element.

Step 5. Review the WBS. Inspect to insure that (1) all PBS elements have nouns (and perhaps an accompanying adjective), (2) all TBS elements have a command verb and object, and (3) all elements have a unique WBS code.

Different Ways to Explode the WBS

Work breakdown structures can be exploded in several ways. One way is to break a product into subproducts, just as has been described in the preceding paragraphs. You list the product and subproducts and the TBS elements to build them. This approach is shown in **Exhibit 6.1-5**.

A second way is to write the overall product at the top. You can the break the WBS into phases. Beneath the applicable phase, you should record the subproduct(s) resulting from each phase. After that, you develop the TBS portion. This approach is shown in **Exhibit 6.1-6**.

A third way is to write the overall product at the top then break the WBS down by areas of responsibility, such as marketing, accounting, etc. Beneath each applicable area of responsibility record the subproduct(s) produced—this may entail several levels. After doing that, you develop the TBS portion. This approach is shown in **Exhibit 6.1-7**.

These are the most common ways to explode a WBS. However, there are no hard and fast rules. You want to build a WBS to meet your needs—develop one that helps you to lead, define, plan, organize, control, and close your

Methods for Displaying the WBS

Just as different ways exist for exploding a WBS, you can use several methods for displaying it.

One popular method is to place a large blank white sheet of paper on a wall. Using Post-its, you can build the WBS. Each Post-it represents one element in the WBS and has a written description on it. If an element seems out of place in one area of the WBS, you simply remove the Post-it and place it where it does fit. This method allows great flexibility in thinking and enables easy modification of the WBS.

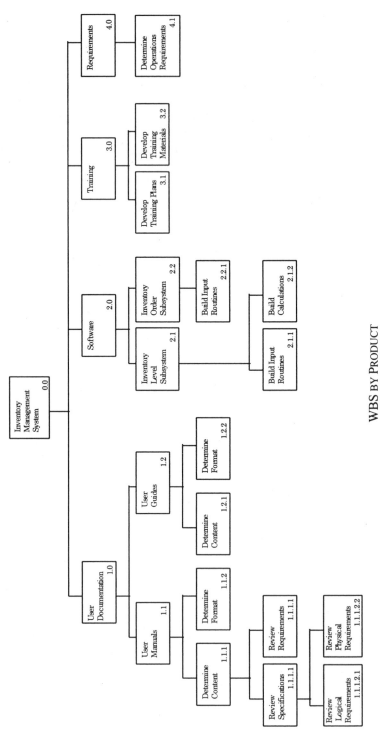

WBS BY PRODUCT
EXHIBIT 6.1-5

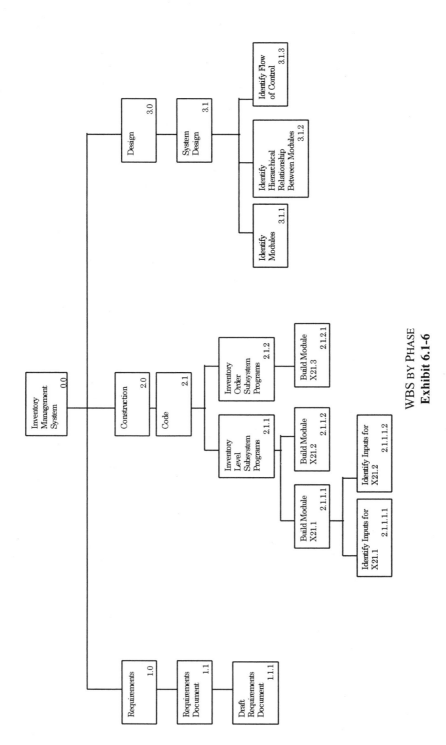

WBS BY PHASE
Exhibit 6.1-6

79

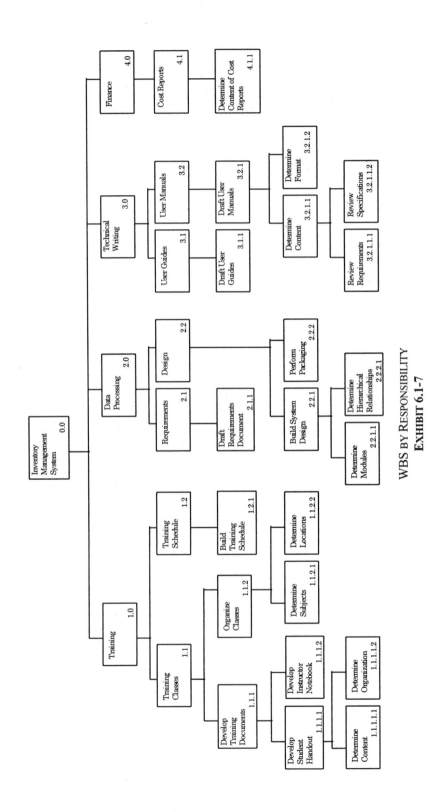

WBS BY RESPONSIBILITY

EXHIBIT 6.1-7

Another method is to draw the WBS on a large whiteboard. On it, you draft different versions of the WBS until you get it right. Then, you transcribe it to paper. Like the previous method, it allows great flexibility in thinking and enables modification of the WBS.

Still another method is to draw it on paper. This traditional method works but can prove cumbersome after awhile, especially for large projects involving hundreds, even thousands, of WBS elements. Little "holes" resulting from erasures make the task grueling.

If possible, you might consider yet another method. Using a graphics package on a personal computer along with a printer plotter, you can draft different versions of the WBS and archive each one. After each review, you make the necessary modifications until you receive everyone's concurrence and produce the final version.

Regardless of method, you will need to transcribe the graphic version of the WBS to paper, usually in outline form. Typically, the paper version has the following format. Each element within the WBS shows the first WBS number and then the verbal description (nouns for the PBS elements and command verbs plus objects for the TBS ones). Each one is indented under the next highest level element, thereby reflecting its level within the WBS. An excellent medium to record this information is on a personal computer using a project management, spreadsheet, or word processing software package. Whenever a modification is necessary, you simply make the change and print the new version.

Sources of Information for the WBS

The act of actually building a WBS is time consuming and frustrating. Despite the effort required, the WBS is absolutely essential to leading, planning, organizing, controlling, and closing a project. But building the WBS may be the easy part. Getting the information you need and the cooperation you require may be far more difficult.

Basically, you have two sources of information. One is existing documentation; the other is people associated directly or remotely with the project.

Documentation can be divided into two groups: documentation that is related to the project and that which is not.

In the first category are documents that specifically describe the scope of the project, its deliverables (or products), and other pertinent factors. Typically, the statement of work is a major source of information. Studies and reports developed prior to the statement of work, such as a feasibility study, may give you additional information.

In the second category are documents that can give you excellent clues as to what tasks will inevitably be conducted and for what subproducts. Organization charts, policy statements, operating procedures, and work measurement studies are just some of the documentation to reference.

Documentation is not the only source of information. Indeed, it serves more as a supplemental source rather than a primary one. People are your primary source. However, not just anybody can be a primary source of information. You should contact persons who have direct, or even indirect, relationship to the project. Even though people with only a remote relationship to the project may not participate in building the work breakdown structure, sometime during the project they may object to the WBS. The principal objection may take the form: "Hey, I would participate in this task in any way. But you did not ask for my input. Therefore, I disagree with what you have."

Still another significant reason exists for soliciting the participation of people only indirectly involved. To be thorough, you need to acquire as much detailed information as possible. Without receiving information from everyone, directly or remotely related to your project, the quality of the WBS suffers. This can lead to missed tasks that can hinder the validity and usefulness of your schedules.

Consensus Attainment

Once you receive the participation of all direct and indirect parties, you should solicit each person's concurrence and have them sign their portion of the WBS. When they sign off, they are agreeing to abide by the breakdown of elements.

Before you receive signoffs, prepare the WBS in draft form. After giving all persons a copy and allowing them sufficient time to review it, hold a review session with them as a group. Solicit their feedback and incorporate their changes. Then repeat everything until you receive unanimous approval. If you fail to receive approval on any portion, the chance of facing considerable resistance later during the project increases

Qualities of a Good WBS

How do you know when you have a good, useful WBS? At a minimum, your WBS should have several qualities. These qualities include the following:

- All legs of the WBS should be factored down to the lowest level, keeping in mind the 80-hour rule.
- All the WBS elements should be uniquely numbered.
- All PBS elements should be written as an adjective and a noun.
- All TBS elements should be written as a command verb and an object.

- All tasks in the WBS, no matter how insignificant or remote, should be identified.
- Feedback and approvals should be obtained from everyone in their respective areas affected by the WBS.

Receiving everyone's approval on the WBS does not mean that it is immune from change. As the project progresses, you will need to update it by adding, amending, or deleting items. Any changes to the WBS should go through the toughest scrutiny. You should have effective change control procedures in place which entail weighing the impact of a change and making an appropriate "go" or "no go" decision. By no means should you take changes lightly, because they will impact how you can proceed and progress on the project. Change can affect your schedule and cost estimates.

Version Control of the WBS

A smart project manager maintains control of all versions of the WBS. You should never discard the previous version simply to maintain traceability in case something troublesome arises due to the change. You may even elect to return to your original plans. Version control entails dating each WBS printout and giving it a version number.

An effective and efficient way to maintain version control is to store the WBS on an automated system using a project management, spreadsheet, or word processing program. When you add or delete an item, you simply redo the WBS numbers, if desirable. Some application programs will even automatically adjust the WBS numbers for you.

Some companies have developed "model" work breakdown structures. These are generic for use on all projects of similar size and complexity. You modify the WBS to meet the needs of your project. These modifications may include adding, modifying, or deleting elements.

The advantages of a generic WBS are twofold. (1) It reduces the amount of planning time required for a project since the laborious task of building a WBS is greatly reduced, and (2) it reduces planning costs since fewer labor resources are required to develop a WBS. It also serves as a baseline from which meaningful discussions can arise since it enables people to talk about issues they would have, or preferred to have, ignored.

In some companies, project managers develop a WBS dictionary. This is a listing of all WBS numbers, their accompanying descriptions, and the output resulting from each WBS element. You may wish to utilize this strategy if you require this level of detail. Just a listing of the WBS number and an accompanying description will usually suffice.

Conclusion

The work breakdown structure is one of your most important tools for leading, defining, planning, organizing, controlling, and closing a project. It serves as the basis for making time estimates, constructing schedules, developing budget projections, allocating resources, assigning responsibilities, and measuring progress. You must make a concerted effort, therefore, to develop a meaningful, useful work breakdown structure.

30100 WORK BREAKDOWN STRUCTURE

P² M² CYCLE — INPUTS · MEASURES OF SUCCESS · TASKS · RESPONSIBILITIES · OUTPUTS

Sources of Information	Statement of Work	Major Participants	TASKS	Project Manager	Senior Management	Project Sponsor	Project Team	Client	Procedures	Work Breakdown Structure
■	■		30105 Determine the overall approach to the work breakdown structure	■						■
■	■		30110 Obtain various sources for the work breakdown structure	■						■
		■	30115 Assign responsibilities for building the work breakdown structure	■		■	■	■		■
■	■		30120 Explode the product to various levels of sub-products	■		■				■
■	■		30125 Determine the tasks and sub-tasks of the task breakdown structure until reaching the weak package level	■			■	■		■
■	■		30130 Provide command verb description and code for each task within the WBS	■			■	■		■
■	■		30135 Apply the 80 hour rule	■			■	■		■
		■	30140 Obtain consensus	■			■	■		■
		■	30145 Determine who will review the WBS	■			■	■	■	■
		■	30150 Determine who must approve the WBS	■	■				■	■
		■	30155 Determine who should receive a copy or selected portions	■					■	■
		■	30160 Institute version control	■					■	■
		■	30165 Establish a revision process	■					■	■
	■	■	30170 Determine how to display the WBS code	■					■	■
			MEASURES OF SUCCESS							
			Is the WBS exploded to sufficient detail?							■
			Have all the major participants given their concurrence to the WBS?							■
			Is the WBS constructed in a consistent, clear and readable format?							■
			Does the WBS identify all the sub-products, tasks and sub-tasks needed to accomplish the goals and objectives of the project?							■
			Does the WBS incorporate all the major deliverables and responsibilities listed in the SOW?							■
			Is the WBS placed under configuration control?						■	■

6.2 ESTIMATING (30200)

The completed work breakdown structure provides you with the starting point for the difficult work of **estimating** the time required to complete each task. Estimating is difficult because so many factors enter into your thinking that impact your accuracy.

Initially you will estimate not the time required to complete the project but rather the time required to complete each element at the work package level within the WBS. The work package level represents the lowest-level elements that are shown on each leg within the WBS. The summation of all of estimates at the work package level gives you the total time required to complete the project.

Reasons for Estimating

Estimating gives you an idea of the time required to complete each task and the entire project. Without such information, you would be unable to determine exactly when tasks and the entire project must finish.

Estimates also allow you to determine, in advance, the level of resources necessary to complete a task and the entire project. With reliable estimates, you can ascertain, for instance, the human resources and equipment required to complete a task.

Estimates play another important role. They enable you to determine which activities are critical and which are not. Because you will use estimates to build a schedule, you can determine which tasks are more important than others; this is known as your critical path.

You can also use estimates to measure performance. If you have reliable estimates (i.e., having a high degree of accuracy), you can use them as standards to determine how well project team members have performed. The estimates can serve as "yardsticks" to assess employee performance. Hence, performance evaluation becomes less subjective.

In addition, you can use those very same estimates to assess how well the project is progressing. With accurate estimates, you track and monitor by noting any variances (variance = planned – actual). If you detect a negative variance, you can respond accordingly.

Finally, estimates are indispensable for developing a schedule. You must have time estimates to build a schedule.

The Politics of Estimating

To realize the benefits of estimating, you must remain conscious of the "politics" of estimating. The opportunity of others to degrade the reliability and validity of your estimates is constantly present.

Some people are afraid to make estimates. They may fear reprisal from management for making an inaccurate estimate. They may not want to go through a feeling of embarrassment if they exceed the estimate. They may fear over-committing themselves. When you force them to make an estimate, you receive estimates filled with qualifications making the estimates virtually meaningless.

Other people do not hesitate to make an estimate. To them, estimating is like making a wager. They will gladly give you an estimate without seriously considering or having appreciation of the impact of their estimate. Their estimates are often meaningless simply because little reasoning power has gone into their formulation. This type of estimating is akin to betting on a horse at the races because the animal has black hair and a white tail. The estimate is whimsical.

Still other people give inflated estimates. Their motive is sometimes sinister. By inflating the estimate for a task, they can complete it in less time. They then achieve the result they wanted. They look artificially effective and efficient when their performance is the result of a bad estimate and not because they are good performers.

Not all inflated estimates have a sinister motive behind them. Some people give inflated time estimates because they are pessimists. They consistently see only the negative. Although appearing harmless, such estimating attitudes can lead to waste. Project managers may allocate more funds than necessary or purchase more supplies than are needed.

Estimates can also be deflated. Sometimes this is an expression of extreme optimism and positive attitudes. But this can prove dangerous and translate into problems as soon as a task goes awry. Such estimating does not account for the possibility that something may go wrong. This can cause embarrassment when you tell the client you cannot meet a certain date and may require additional time. It can also cause you to use scarce time to replan.

Some people agree to a bad estimate because they have been pressured into it by someone with higher authority. Perhaps the boss of a project manager has agreed to accept a bad estimate because agreement was politically advantageous. The boss then pressures the project manager into to accepting the bad estimate. Such estimates deceive everyone, but only for a short time. Once the project begins, the project manager will see and feel the impact of the poor estimate.

Obstacles to Acquiring Good Estimates

There are many obstacles to acquiring good estimates. Lack of available information is one of the most common obstacles. You might need information about a specific task, such as the skills and the accompanying level of expertise to perform it. Such information is often not readily available. Or you may need to conduct further research, which may require considerable time (which may not be available).

Even if you acquire an estimate, you may face another obstacle. The estimate that you receive may be less than genuine. It could be under- or overinflated, causing serious skewness in your estimate. Detecting this under- or overinflation is not easy. You have two ways to overcome the problem. One solution is to adjust the estimate accordingly to compensate for the skewness. A second solution is to contact someone else to verify the estimate for accuracy, such as consulting someone else for another opinion.

Management intervention is another common obstacle to making accurate estimates. Sometimes management will place pressure on a project manager to take only a specified amount of time to complete one or more tasks. Or management may pressure the project manager to derive an estimate too prematurely. In both cases, degradation of reliability and validity of the estimate may occur.

Lack of time is another obstacle project managers may face. They may not have the time to derive good estimates simply because management will not give them enough time. Consequently, project managers yield, mainly because of career interests, to the pressure and hope that all works out for the better. If project managers do not express their need for more time, they face even worse pressure when they concede to unrealistic estimates that they will not be able to meet later in the project.

A dearth of money is another obstacle. Often funding is not available for the resources, such as labor, to develop good estimates. To save money in the short term, project managers reduce the time for estimating only to find later they must request more time because their estimate was no good. The result is embarrassment for the project manager and an angry management or client or both.

If these obstacles were not enough, many project managers lack cooperation from the client, their management, and their project team. For political reasons, management may not want to provide time or resources to do estimating. The client may not give you any information. Your team members may not like working on the project and will do anything to hide information that will improve your estimates. Such circumstances can lead to early project failure simply because this lack of cooperation may spread to other aspects of the project.

Closely related to lack of cooperation is lack of communication. Project managers may face insurmountable communication problems that prevent access to important information for making good estimates. Distance often contributes to this obstacle. Team members may be located throughout a wide geographical area; project managers may have to make estimates on behalf of some members. Clients may not always be accessible, and project managers might have to guess at the effort required to perform a specific task.

Poor formulation of goals, objectives, and requirements are additional obstacles that project managers sometimes face. Occasionally, people will be assigned as project managers to a venture that has an ill-defined purpose. In this circumstance, estimating is extremely difficult. Project managers will be unable to determine exactly how long it will take to complete the project if they lack a solid idea of what will result from the project. Under some circumstances, the objectives and requirements are vague, too, leading to "guesstimates" rather than estimates.

Poor formulation of goals, objectives, and requirements can lead to an inadequate WBS—this deficiency is a proven potent obstacle to good estimating. Without a good WBS, project managers will find it very difficult to determine specifically the time required to perform a given task and the entire project. A cornerstone for a useful work breakdown structure is having well-defined goals, objectives, and requirements. Unless you have those, the WBS will be inadequate, leading to unreliable estimates.

Lack of knowledge of common estimating techniques is another obstacle that project managers may face. Frequently, project managers do not have any understanding of the requirements for developing useful estimates. About the only method they know is asking someone for an estimate and using that figure. Alternatively, they may develop an estimate based upon their own knowledge and experience.

Warnings About Estimating

Despite all the obstacles, you can perform good estimating. Doing so, however, requires remembering some important thoughts.

You must place parameters on your estimates. For the estimates to have some semblance of accuracy, you need to assess the level of variability between what you consider is an accurate estimate and an inaccurate one. You must have a way to compensate for the tendency towards having under- and overinflated estimates.

You will never have a completely accurate estimate. You can only strive towards accuracy. You cannot account for all the factors that impact the validity of your estimates. The best you can achieve is a "reasonable" confidence level. Occasionally, something occurs that invalidates your estimate for a task. This is common and may require replanning.

Your estimates will sometimes have to change because of varying conditions. You may not have expected technical complications with your equipment or political disagreements among project participants. These and other circumstances will force you to reconsider your estimates and make new ones. Replanning is absolutely necessary, or your result will be an unrealistic schedule.

Estimating involves some level of subjectivity. No matter who makes the estimates, including yourself, you will incorporate some attitudes or feelings to determine the time required to complete a task. These attitudes will skew your notion of the time required to complete a task. For instance, someone else may estimate the time required to perform the same task, and that person's estimate may be substantially different from your own.

Estimating involves considering an indeterminate multitude of variables. No matter how hard you try, you can never account for all of the aspects that will impact the validity of your estimate. You will miss something. The idea is to attempt to identify as many of the variables as possible that go into an estimate.

You should never blindly accept an estimate from someone. Such estimates may differ substantially from your own. Even if they may match, the basis for the estimate may differ dramatically from your own. Later in the project, some of the variables you or the other person never considered may arise, thereby negating the estimate.

Carefully choose who will give the estimate. Do not select just anyone unless you are willing to settle for "just any" estimate. How do you know whom to contact for an estimate? You can request people doing the work to make an estimate. You can contact those people who participated in the development of the WBS. Or you can solicit estimates from people who will be impacted remotely by some tasks in the WBS. These people may not necessarily perform the tasks, but their cooperation is necessary to complete the project.

Document all estimates. When someone says a task will require 70 hours of work, have them document it in a memo. If they fail to do so, you should draft a memo of understanding to record that the person will work on a task or set of tasks for a specific amount of time. Be sure to give the estimator a copy of the memo. A wise action is to send copies to key project participants. Documenting estimates breeds commitment and accountability. The impact is mainly psychological. When someone sees their name visible before the team, they find it very difficult to back down on their promise.

Check whether historical records of previous projects exist. These records will help you verify the accuracy of your current estimates and save you from "re-inventing the wheel." The past, if well-documented, can save you from facing repeatable future problems. Unfortunately, good records do not always exist or they may be inaccessible.

Ensure that you have a detailed work breakdown structure, as this is your best weapon for devising worthwhile, useable estimates. If the level of detail is sufficient, you have something tangible and measurable. You cannot use a broad WBS element to measure progress, because it covers too much. If a WBS element is specific, you will then have something that is useful.

You can see that calculating a precise estimate—a contradiction—is futile. About the best you can do is to approximate one—hence the reason for calling it an estimate. Making a time estimate is similar to Kentucky windage. You can take an approximate aim and fire, hoping to come close to hitting the target. Or you can take the sometimes popular "ready, fire, aim" approach, and pray for success.

Steps in Estimating

Ironically, the steps in estimating are simple and straightforward. They are twofold: first, identify the lower-level items in the WBS, called the work package level. This level contains the elements that are easier to use as a "device" to track progress than at the higher levels in the WBS. Second, estimate the time required to complete each element. You have several techniques at your disposal to do estimating. These are the unscientific estimate, the PERT estimate, the global efficiency factor, and the productivity adjustment effort.

Unscientific Estimating

This technique is known by other less flattering terms, such as a ballpark estimate, a "shot in the dark estimate," or SWAG (scientific wildly assumed guess). Although this technique is fast and easy to use for estimating, its results lack reliability.

The technique is highly subjective. Someone makes an estimate based on the requirements to perform a task. You may, for instance, ask two people with equivalent skills and experience the time needed to perform a task. The amount may differ dramatically perhaps because each person has a different perspective. If the estimator has no real knowledge of the task, there is even more danger of understating or inflating the estimate.

The unscientific technique also lacks parameters. The estimator accounts for no variability in the estimate. No minimum or maximum value is cited or consciously included in the estimate. The estimator gives little or no consideration for the best and worst conditions. The unscientific estimate is not recommended for projects unless the project uses experienced, highly skilled employees having little or no history of turnover and the project itself is well defined and requiring little innovation. If these conditions do not exist, you should consider unscientific efforts with extreme caution.

PERT Estimate

Project managers employ this technique when precise time estimates are not feasible. Typically, these are projects requiring innovation and for which the functionality of the product is more important than finishing a project on time. Not surprisingly, the PERT technique has its origins in research and development environments.

PERT Formula

The technique requires making three time estimates per task and then combining those estimates to derive a single figure.

- The **most likely (ML) estimate** represents the time required to complete a task under "normal" or "reasonable" conditions. You might consider, when making the estimate, what is the "typical" time needed to do a task.
- The **most optimistic (MO) estimate** represents the time required to complete a task under the "best" or "ideal" conditions (i.e., if no obstacles, such as lack of information or cooperation, confront you.) It reflects the least amount of time needed to perform a task.
- The **most pessimistic (MP) estimate** represents the time required to complete a task under the "worst" or "nightmarish" conditions (i.e., every conceivable obstacle confronts you). It reflects the longest amount of time needed to perform a task.

You perform this technique for each work package level (the lowest level elements on each leg of the WBS). Having these three estimates, you can "plug" them into this formula:

$$\text{Expected Time} = \frac{MO + 4(ML) + MP}{6}$$

where

ET = expected time
MO = most optimistic

ML = most likely
MP = most pessimistic

Example of PERT

The following example shows a marketing proposal. The work package level tasks and the estimates (in hours) might look like this:

Task	MO	ML	MP	ET
Determine contents	8	32	50	31.0
Determine format	2	8	24	9.7
Determine printing requirements	2	8	12	7.7
Develop draft	16	40	80	42.7
Determine distribution	2	10	14	9.3
Determine word processing support	1	3	7	3.3
Totals:	31	101	187	103.7

Your job, however, is not complete. You must also adjust the expected time by a percentage to account for lost time and interruptions. A typical amount is 7 to 10 percent to account for this loss of productivity.

Task	ET	Adjustment Percent	Adjusted ET
Determine contents	31.0	1.10	34.1
Determine format	9.7	1.10	10.7
Determine printing requirements	7.7	1.10	8.5
Develop draft	42.7	1.10	47.0
Determine distribution	9.3	1.10	10.2
Determine word processing support	3.3	1.10	3.6
Totals:	103.7		114.1

After calculating the total time for each task, you can then quite easily determine the amount of effort necessary to conduct the entire project.

You can use the form shown in **Exhibit 6.2-1** to calculate and record your estimates for each activity and the total project. Just follow the instructions for the exhibit.

Activity Number	Activity Description	Most Optimistic	Most Likely	Most Pessimistic	Expected Time	Productivity Adjustment Factor	Revised Expected Time	Clock Hour Divisor	Duration
[1]	[2]	[3]	[4]	[5]	[6]	[7]	[8]	[9]	[10]
2.1.3	Determine customer needs	10	12	22	13.3	10%	14.7	8	1.8
2.1.4	Review with team	4	7	9	6.8	20%	8.2	8	1.0

ACTIVITY ESTIMATING FORM

EXHIBIT 6.2-1

To complete the form, follow the instructions below by matching the applicable number with the corresponding one located in Exhibit 6.2-1:

[1]   Numeric designation uniquely identifying the activity
[2]   Short narrative description of the activity
[3]   The total time to complete the activity under ideal conditions
[4]   The total amount of time to complete the activity under normal, or average, conditions
[5]   The total time to complete the activity under the worst conditions
[6]   The expected time to complete the activity using the PERT formula
[7]   The percentage nonproductive time employees expend while working on the activity
[8]   The percent in column [7] multiplied by the figure in column [6]
[9]   The number of hours per day that employees will work on the activity
[10]   The number of days the activity will take to finish; derived by dividing the figure in column [9] into the one in column [8]

ACTIVITY ESTIMATING INSTRUCTIONS
FOR COMPLETING THE FORM SHOWN IN EXHIBIT 6.2-1

Advantages of PERT

The PERT technique offers several advantages. One advantage is that it places parameters on estimates, forcing you to consider both understated and inflated estimates with a third estimate, the most likely estimate. The result is a value balanced between the two extremes, thereby giving more meaningful and reliable estimates.

The technique forces critical issues to surface early in a project. Typically, estimates do not consider "external" factors that could impact an estimate's validity. For example, approvals from certain people indirectly involved with the project are a factor often overlooked during the beginning of a project. The PERT technique encourages discussion of such matters early rather than later in a project. Thinking about the most pessimistic, most optimistic, and most likely values makes the estimator think about all factors impacting an estimate.

The PERT technique by encouraging discussion about a wide range of issues results in better communication. When implementing the technique, many project managers bring together people in a large room and they discuss the length of time a task will take. These people discuss the issues impacting the estimate. Communication barriers fall and misunderstandings are addressed. All this activity occurs before the project begins and not later (when it may be too late).

Finally, the PERT technique leads to more detailed planning. If project managers feel an estimate is too large, such as exceeding the two-week rule, they can break, or explode, the WBS element into smaller subtasks; each subtask would not exceed more than two weeks of work. Even if the derived estimate does not exceed the two-week rule, project managers can decide to decompose a task into subtasks to have more meaningful elements to track.

Disadvantages of PERT

Despite the advantages of the PERT technique, there are criticisms about its use. The most common one is the time it takes to do a PERT estimate. Project managers and selected project participants must determine the values for the most optimistic, most likely, and most pessimistic estimates. For a project of fewer than 50 tasks, the time required for this technique is not that high. However, projects with more than 50 tasks require considerable time.

The technique can also lead to frustration among project participants. People may argue for hours over the most pessimistic value for a task. This problem may not be as much with the technique as obtaining consensus.

Because the technique requires so much time, people participating in the estimating process are away from production; they are spending time making estimates rather than being "on the floor."

Finally, the technique requires extensive calculations. You have to derive an expected time for each task, which can number in the thousands. The project manager becomes more involved with "number crunching" than with managing the project.

While these criticisms have some merit, they are extremely shortsighted. True, the PERT technique is time consuming but that is an advantage, too. It forces the project participants to plan early; they know not only what to do but the time to perform each task. This planning forces them to direct their efforts efficiently and effectively. They are not given a free rein to do just anything they want for as long as they want. The time spent up-front on a project saves much more time later.

The criticism that the technique leads to frustration is not totally valid. The discourse over the time required to perform a task should lead to greater cooperation. Those persons who feel frustrated as a result of the technique are the ones with the problem. They may lack team-building skills and, by nature, may never agree with anyone on any issue, or it may be that they cannot see an issue from a different perspective. In other words, unsatisfactory results from using the PERT technique may be a consequence of the user rather than a failure of the technique itself.

Concern over removing personnel from production also bears reconsideration. Although people may be removed from production in the short run, these people will be more productive overall because they will know their responsibilities on the project and have more accurate time estimates for their activities—estimates the team will have created together.

The final criticism that the PERT technique involves too many calculations may arise from people's laziness rather than the technique. You can overcome the tedious calculations by utilizing a personal computer spreadsheet program. By entering the most pessimistic, most likely, and most optimistic values for each task, the expected time can be automatically calculated.

Global Efficiency Factor

This technique is simple but extremely subjective. It involves assuming that employees are 100 percent productive. You then deduct from that 100 percent an estimated percentage for each "deficiency" on the project. A deficiency is any shortcoming that exists that will impact productivity.

Hence, let us assume you are the project manager for an office automation team. Your goal is to automate all operations in your department. The first step is to list all the deficiencies and accompanying percentages accounting for the impact of each deficiency.

Deficiency	Percentage
Low morale	15%
Unsatisfactory skill level	5%
Inadequate familiarity with project	10%
Inadequate equipment	5%
Poor specifications	10%
Total Deficiency	45%

You then subtract the total deficiency factor from 100 percent to derive a global efficiency factor (GEF). In this case:

$$100\% - 45\% = 55\%$$

Using the GEF, you divide it into the amount of time that you had originally estimated to perform each task. For instance, you had estimated to select the right personal computer workstations at 150 hours. You divide the 150 hours by 55% to derive the estimate that you would use to determine how long to perform a task:

✓   150 hours divided by 55% = 273 hours

Hence, your original estimate of 150 hours has been adjusted to 273 hours to compensate for the deficiencies.

The most redeeming value of this technique is its speed. You can quickly list each deficiency on the project—at least, that you are aware of—and assign a percentage to reflect the weight of each deficiency.

The problem is that you sacrifice accuracy for speed. Determining a percentage for a deficiency is subjective and may vary from project manager to project manager. Quite often, the GEF is used by project managers for the same reason that they use the unscientific estimates: convenience.

Productivity Adjustment Percentage

This technique is similar to the global efficiency factor. However, the percentage you use has its origins in objective measurement. Most large institutions have conducted industrial engineering studies to determine the overall productivity of their firm. For instance, a firm might have a productivity level of 75 percent, meaning that 25 percent of the employees' time is nonproductive, such as time spent milling about the office coffee pot discussing the scores of the weekend football game.

You will use this productivity factor to adjust your estimates for each task. Using the office automation project again, you originally estimated that selecting the right personal computer workstations would take 150 hours. Your company's overall productivity rate is 75%. You would subtract the 75% from the 100% to obtain your nonproductive time. Then, you would multiply the 150 hours by a factor of 1.25 to derive your productivity adjustment estimate that you will use to calculate your dates. Thus,

100% − 75% = 25%

1.25 × 150 hours = 187.50 hours

This technique has several advantages. First, it is quick for doing calculations after obtaining the original estimates. Second, the percentage has an objective basis, unlike the global efficiency factor.

The technique is susceptible to several criticisms. First, the percentage used for adjustment is global and may not be relevant to certain tasks. Usually, the percentage reflects a mix of blue collar, white collar, and managerial work. Hence, the percentage may be too low or high for a specific type of work, such as information systems or accounting. In other words, the percentage may include a mix of "apples and oranges."

Second, the productivity adjustment estimate may merely authorize more time to perform a task. Rather than allocating 100 hours for a task that could be done in that time, for example, you could adjust it to allow for 125 hours. Hence, you would be allocating more time than is necessary to complete the task, thereby increasing your costs.

When Your Estimates Are Too High

Occasionally, project managers will complete their estimates only to find that the amount of work to develop the product is greater than they had envisioned. The estimates may indicate that the magnitude of effort exceeds what senior management and the client are willing to authorize and may make it impractical to finish by the project completion date. As a project manager, what can you do?

You can verify your estimates by seeking a second opinion. You may find some estimates are too inflated. You can investigate problem areas by talking with people who are not on your project but who have worked on ones similar to your own. You can also look at historical records of previous projects to determine the time it actually took to perform certain tasks.

Another option is to reduce the scope of your project. You can develop a product that is a smaller version of the original design. You would thereby reduce the level of effort.   If you elect this option, be sure you receive the client's acceptance.

Still another option is to reduce the quality of workmanship on the product. This option is fraught with danger. This option could tarnish your reputation. You could produce substandard work that would prove very costly in the future. For example, there is little to be gained from developing a substandard engine that would lead to recalls and endless legal battles.

You can also use output from previous projects of a similar nature. Why bother to repeat something? You can save time and money if you start by duplicating earlier material. True, you may need time to modify the material but the total amount of time required should be less.

There are two warnings about using material from previous projects. First,, sometimes using another project's material may take longer. You will need to review and understand the material. This could take a long time, based on the complexity of the material and magnitude of the necessary adjustments. Second, every project is different. The applicability of material from one project to another may not work, even though substantial similarity exists.

Acquiring more productive people is an effective way to lower time estimates. You should acquire people with top-notch qualifications, much higher than you would ordinarily hire on a typical project. These people should have greater experience and skill than you would require under normal conditions. By taking this approach, however, you face the possibility of higher labor costs to the point that your budget projections may become invalid.

Another possible consequence is that qualified people may eventually feel unchallenged, thereby leading to lower productivity and higher turnover. As a project manager, you must remain conscious of this problem. An effective way to prevent its occurrence is to provide opportunities for growth and employee participation in the decision-making process.

Another option project managers seldom employ is streamlining operations. Few realize the impact that the number of forms or the complexity of reporting procedures can have on the time required to complete a task or an entire project. By consolidating forms and reducing the number of approvals, you can reduce the flow-time of your project.

This option does have a serious repercussion. If you streamline operations, you alter the status quo. The change can disrupt operations and upset people, resulting in lower productivity and sliding schedule.

Finally, you can acquire better equipment or facilities. Improvements in these two areas can dramatically improve productivity. New, state-of-the-art machinery can increase output, as can increasing the number of work units per employee.

Improvements in equipment and facilities must occur in a planned way if you want to maximize productivity. Unless you provide time for training, employees will feel frustrated and productivity may decline. In addition, providing new facilities requires time for people to adjust to their new surroundings. Failure to give them time can lead to poor productivity and may extend the period for completing a task or the entire project.

When Your Estimates Are Too Low

From time to time, you may suspect that estimates for specific tasks, and not necessarily the entire project, are too low. These low estimates are frequently the result of optimism by the estimator. If you face this situation, you have some ways to increase your estimates.

You can verify your estimates for accuracy by seeking a second opinion; asking a third party the time required to complete a task, or checking the historical statistics on projects with a similar task. The statistics can give you a realistic appraisal on the time required to complete a task.

You can also pad all estimates by a certain percentage. This technique allows you to "compensate" for the level of optimism your original estimates may incorporate. You can use a percentage across the board (a flat percentage applied to all tasks) or a variable percentage (a percentage contingent upon the specific activity in question).

Finally, you can challenge the original estimators by asking them to sign a document indicating their commitment to comply with their estimate. Once you approach them for their signatures, the chances increase for obtaining a realistic estimate.

Steps to Acquire Estimates

Acquiring your estimates is a simple 10-step process. **Exhibit 6.2-2** illustrates the sequence of these steps.

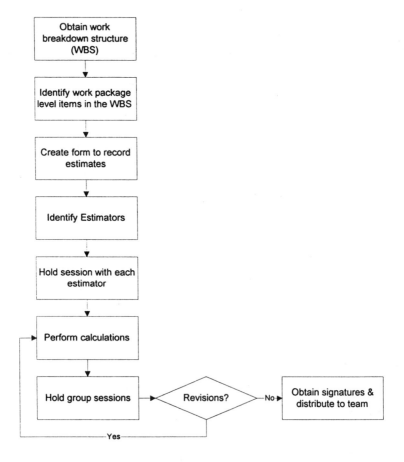

STEPS TO ACQUIRE ESTIMATES FOR TASKS IN THE WBS
EXHIBIT 6.2-2

1. Obtain a complete work breakdown structure. Make sure that the WBS is final. You cannot do your estimates without this document, because it gives you the tasks needed to manage your project.
2. Identify all the work package level items in the WBS. These items are the lowest ones in each leg of the WBS. You will make estimates for each one.
3. Develop a form, like the one shown in **Exhibit 6.2-3**, to record your estimates. This will guide you through the estimating process.

Activity Number	Activity Description	Duration (Days)	Total Hours
[1]	[2]	[3]	[4]
3.5.1	Determine format	2	16
3.5.2	Review requirements	3.5	28

[1] Numerical designation uniquely identifying the activity
[2] Short narrative description of the activity
[3] Number of calendar days that the activity in column [1] will last
[4] Cumulative hours required to perform the activity in column [1]

ESTIMATING FORM AND INSTRUCTIONS
EXHIBIT 6.2-3

4. Identify who will make estimates for which tasks. Typically, they should be the people who will do the work, and there may be more than one person per task. If more than one person is involved, be sure to contact all of them. The person designated as prime estimator for the activity reconciles estimate variations.
5. Hold individual sessions with the estimators. Refer to the form shown in **Exhibit 6.2-3** and be sure to record any other useful information arising from the session, particularly any differences of opinion.
6. Perform the calculations. This is especially critical if you use the PERT estimating technique. Regardless of the technique selected, make sure the estimates are displayed clearly.

7. Hold group sessions, with all estimators attending, to resolve differences. These meetings will not only help resolve differences in estimates but will help to open the lines of communication among people working on the same task. You should encourage this discussion and not dismiss the session until everyone agrees on the time required to complete each task.

8. Make the revisions to the estimates and record these changes. Be sure to record the latest revision and archive previous documents. A good practice is to give each person a copy of only those estimates that have been agreed upon.

9. Have everyone sign the document. This action will make estimators take seriously their estimates. Signatures are evidence of their agreement and breeds commitment.

10. Distribute a compilation of documents that everyone has signed. This action gives you the visibility needed to ensure everyone works according to the estimates.

More important than actually following these steps is obtaining accurate estimates and the cooperation of people who will follow them. Getting signatures and visibility are minimum requirements for developing reliable schedules.

Conclusion

Estimating is just what the word implies. When you do it, you are making approximations. Any time you do that, subjectivity is involved. You may estimate using an algorithm, but it is the inputs to the formula that determine the validity of the estimate. What values you assign to the variables in your estimate calculations greatly affect the results.

30200 ESTIMATING

INPUTS					P² M² CYCLE / TASKS	RESPONSIBILITIES					OUTPUTS	
Major Participants	Political Pressures	Work Breakdown Structure	Statement of Work	Risk Identification	TASKS	Project Manager	Senior Management	Project Sponsor	Project Team	Client	Procedures	Time estimates
	▪				**30205** Determine the political pressures that you will face when estimating	▪						▪
	▪				**30210** Determine how you plan to deal with the political pressures	▪						▪
▪					**30215** Assign responsibilities for estimating	▪		▪	▪			▪
▪					**30220** Develop a procedure for performing estimating	▪					▪	▪
▪		▪		▪	**30225** Minimize subjectivity	▪		▪	▪			▪
	▪	▪	▪	▪	**30230** Determine all variables	▪		▪	▪			▪
				▪	**30235** Place parameters on estimates	▪						▪
				▪	**30240** Review historical records	▪		▪	▪			▪
				▪	**30245** Determine approach to estimating	▪		▪	▪			▪
			▪		**30250** Apply against detailed work breakdown structure	▪		▪	▪			▪
		▪	▪	▪	**30255** Document all estimates	▪			▪			▪
▪			▪	▪	**30260** Adjust estimates if too low or high	▪		▪	▪			▪
▪					**30265** Determine who must provide feedback on the estimates	▪		▪	▪	▪	▪	▪
▪					**30270** Determine who must review the estimates	▪	▪		▪	▪	▪	▪
▪					**30275** Determine who must approve the estimates	▪	▪	▪		▪	▪	▪
					MEASURES OF SUCCESS							
					Has the appropriate time estimating technique been selected?							▪
					Have all the participants given their concurrence to the estimates?							▪
					Is the WBS used for time estimates, using tasks and sub-tasks at the work package level?							▪
					Do the time estimates reflect constraints identified in the SOW?							▪
					Do the time estimates give some reasonable confidence in achieving project goals and objectives?							▪
					Are responsibilities for estimating documented?						▪	▪

6.3   RISK CONTROL (30300)

In most organizations, project managers find themselves operating in a dynamic environment and managing projects with limited resources. The number of risks, or potential losses, can be high. These losses often negatively impact the project in terms of schedule, budget, quality, or a combination of these three areas. Project managers want to contain these losses, and one of the best ways to do that is through **risk control**.

Risk control involves identifying the critical activities of the project and the threats (undesirable events) to them; this is called **risk analysis**. Project managers must determine the probability of occurrence (such as low, medium, and high) to those threats and the technical, operational, and economic impact on the organization; this is called **risk assessment**. The project manager can then determine the measures to take to prevent a threat from occurring or to lessen the impact of a threat; this is called **risk management**.

The advantages of risk control are many. By identifying the critical activities and the threats to them, project managers can revise their time and cost estimates accordingly. They can focus their efforts on the critical activities to ensure minimal impact on the project. They can derive an overall risk profile for the project and, consequently, prepare themselves for a somewhat uncertain future. Finally, they can use risk control to develop reliable, useful contingency plans.

When performing risk control, you should encourage the participation of all key participants on the project. Involvement of the right people can mean the difference between reliable and useless risk control. Their omission may mean the failure to identify critical activities and the respective threats.

You should also define the most important activities and then identify all the potential threats to those activities. The activities may or may not be the ones on the critical path. This analysis may result in the project manager revising the logic in the schedule to reflect the importance of an activity.

The project manager should use a risk control log, shown in **Exhibit 6.3-1**, to record:

> Critical activities
> Impacts
> Probability
> Threats

A **Risk control log** should be used to develop contingency plans for dealing with each threat to a critical activity. These plans should reduce or eliminate the impact of a risk to a critical activity.

Risk control does not eliminate risks; it simply helps project managers to identify the most important risks to the most important activities. Project managers can never identify all risks. Essentially, all they can do is minimize the number and impact of risks.

The degree of risk control depends a great deal on the amount of time and effort project managers are willing to spend doing it. They may want to immerse themselves in probability statistics and calculating the financial impact of a loss. The extent of risk control should depend on the scale and importance of the project. Obviously, the larger and more important the project, the greater the need for project managers to do risk control as scientifically as possible.

Critical Activity	Threat(s)	Impact(s) [H, M, L]	Probability
[1]	[2]	[3]	[4]

RISK CONTROL LOG
EXHIBIT 6.3-1

You can use the form in **Exhibit 6.3-1** to perform risk control. The appropriate entries are identified below:

[1] The name of the task to perform
[2] The actions that can hinder or stop the progression of the task
[3] The degree to which the threat affects the task
[4] The likelihood of occurrence

Project managers should also utilize the **risk assessment checklist**, shown in **Exhibit 6.3-2**. The checklist can assist in identifying risks for your project.

Risk Assessment Checklist

1. Identify planning risks, related to:
- Statement of work
- Work breakdown structure
- Time estimates
- Budget estimates
- Scheduling
- Product definition
- Automated project management
- Project plan
- Life cycle

2. Identify organizing risks related to:
- Task assignments
- Staffing
- Training
- Project handbook
- Reports
- Forms
- Resource allocation, regarding:
-      People
-      Hardware
-      Software
-      Data
-      Supplies
-      Facilities
- Organization chart
- Client participation
- Senior management support

RISK ASSESSMENT CHECKLIST
EXHIBIT 6.3-2

3. **Identify controlling risks related to:**
- Contingency plans
- Tracking of plans versus actuals, such as:
-     Cost
-     Schedule
-     Quality
- Meetings, such as:
-     Status review
-     Checkpoint review
-     Staff
- Change control
- Configuration management
- Quality assurance
- Milestone benchmarks

4. **Identify technical risks related to:**
- Testing procedures
- Development of life cycle deliverables, such as:
-     Feasibility study document
-     Requirements definition document
-     Alternatives analysis document
-     Functional specifications document
-     Preliminary design document
-     Verification / validation plan
-     Installation and implementation plan
- Control of data
- Development tools
- Development techniques
- User documentation
- Product quality
- Training
- System security
- Operations manual
- Design, regarding
-     Programs
-     Data
-     Procedures
-     Security
-     Documentation

30300 RISK CONTROL

INPUTS				P² M² CYCLE	RESPONSIBILITIES					OUTPUTS			
Potential Risks	Management Direction	Statement of Work	Major Participants		Project Manager	Senior Management	Project Sponsor	Project Team	Client	Risk Management	Risk Assessment	Procedures	Risk Identification
				TASKS									
■			■	**30305** Establish a procedure	■							■	
		■		**30310** Assign responsibilities	■			■	■			■	
	■	■		**30315** Determine when to conduct risk assessment	■			■	■			■	
■		■	■	**30320** Conduct risk analysis	■			■	■				■
■			■	**30325** Conduct risk assessment	■			■	■		■		
■	■	■	■	**30330** Conduct risk management	■			■	■	■			
				MEASURES OF SUCCESS									
				Does the risk control activity result in identifying the major threats to the project?									■
				Does the risk control activity provide reasonable confidence in achieving project goals and objectives?								■	
				Have all the major participants been involved in the risk control activity and concur with the result?								■	
				Is the result of the risk control activity incorporated in the budget time estimates and the schedule?						■			
				Does the risk control activity address any risks identified in the statement of work?									■
				Have all risk management actions been identified to address major risks or threats?						■			

6.4 SCHEDULES (30400)

The work breakdown structure tells you only what to do. It does not give you other information to help you effectively lead, define, plan, organize, control, and close your project. The main instrument to help you accomplish this is the project **schedule**.

Purpose of a Schedule

As a project manager, you need to know the logical sequence of each task and the start and stop dates for each one. The schedule helps you to do derive these dates.

For example, you might want to plan for a project to produce a marketing proposal. You will need to know the steps to complete the project. You will also need to know the logical sequence of each step (i.e., which step goes first, second, third, in parallel, etc.). Once you have identified the relationships among the tasks, you can estimate the time to perform each one, then calculate the start and finish times.

Benefits of Schedules

A schedule identifies when an activity starts and finishes. It lets you plan when the entire project can start and finish, especially if these dates are not imposed by another party. Management should not specify when the project must end; an accurate schedule should tell you when the project can finish.

A schedule offers you another benefit. It tells you the logical sequence of a project. Every task within a project, if managed well, will fall into a sequence, with some tasks preceding and others following. A schedule enables you to follow the sequence you develop. Once your project begins, everything should occur according to the sequence.

Frequently, project managers who are not using a schedule fall into the trap of performing tasks prematurely—before they have received input from another related task. The result is lower quality, poor productivity, a high level of employee frustration, extended project end date, and cost overruns.

Putting a schedule together is not easy. You will need the participation of many people and will require from them not only their information but their concurrence. The very act of creating a schedule forces you to determine what tasks to complete in order for the next ones to occur and when they must occur.

The schedule also provides better control. You can track and monitor progress during the project—knowing which tasks are 100 percent complete; which ones are partially complete; and which ones have yet to start.

Schedules impose discipline on a project. By following the schedule, you force the project to proceed in a specific way and according to the set timetable. Once you detect any deviations from these parameters, you can respond accordingly.

But this discipline goes one step further. It imposes discipline on the people on the project team. Rather than having everyone perform whimsically without regard to others, everyone realizes that their output is another team members' input. This helps to create an atmosphere of teamwork. People realize the significance of their work, understanding the impact they will have on the performance of others. In addition, you, the project manager, can more easily see that impact on the entire project. Armed with that information, you can take appropriate action.

A schedule indicates when you need specific resources. Knowing when an activity occurs allows you to acquire resources in advance rather wait until after an activity has started.

Some project managers fail to identify when they need people with the right skills until after a task has started. Meanwhile, they frantically try to find people with the right skills, and their projects fall further behind schedule. With an accurate schedule, they can avoid this problem by identifying in advance when they need a person with the requisite skills.

Finally, a schedule helps you to determine which tasks are critical and which ones are not. It helps you to identify which activities should be addressed immediately as opposed to trying to do all of them at once. Attempting the latter will only drain your resources (time, labor, money, equipment, etc.), which will tend to compromise quality, lower productivity, delay schedules, and cause budget overruns.

Why Some Project Managers Do Not Develop Schedules

Despite the important advantages to having useful, accurate schedules, many project managers do not make the effort to develop them. Several reasons exist for this attitude.

Some project managers are simply lazy. They do not want to take the time or expend the effort to build a good schedule. Instead, they want to start the project immediately. This is indicative of poor management, shortsightedness, and "tunnel vision." More often than not, their projects fail or "succeed" only after exceeding the original resource estimates and budget.

The best cure for overcoming laziness is for upper management to make it mandatory to have schedules developed before any significant work on a project can begin. Likewise, upper management should make developing good schedules a standard to evaluate performance during periodic performance appraisals.

Another reason that many project managers fail to develop good schedules is their lack of skills. They do not have the proper training or any idea what goes into a developing a schedule.

You can easily overcome this problem by training to develop different types of schedules. You can also acquire skilled resources to perform the scheduling function.

Some project managers fail to develop detailed schedules because they perceive that they lack the time. Of course, individuals who make this claim imply they cannot manage their own workload. One wonders how can they manage their projects if they are unable to manage their own time! Project managers who take time to develop good schedules in the beginning save themselves time later in the project, allowing them to concentrate on important matters.

You can overcome lack of time by prioritizing your workload to eliminate unimportant tasks that consume a large percentage of your time, or you can acquire resources to help with the scheduling. You may also be able to make more time available by removing yourself from your regular work environment with all its inherent interruptions.

Another reason for failure to develop schedules is lack of cooperation from participants on the project. Project team members may lack the willingness to give time estimates or dates to complete tasks for fear of committing themselves. Senior management may waiver over when the project should begin or end, thereby making it difficult to create a schedule. Or the customer may refuse, for whatever reason, to provide information or dates necessary to construct schedules.

You can gain cooperation by documenting your needs in a memo and circulating a copy to selected parties related to the project. Documentation protects you but, more importantly, it also psychologically persuades un-cooperative parties to comply with your request. Making your concerns visible prompts people to respond rather quickly after seeing their name in print. Still another effective way to overcome this problem is to hold a series of sessions with everyone involved on the project, beginning with each individual and then meeting as a group. That way everyone provides the feedback necessary to develop comprehensive schedules and remove disagreements among all the parties.

Lack of management support is another reason for poor schedules. If management does not provide the time or money to develop good plans, your options are limited. Senior management must provide the opportunity to develop sound schedules. The best way to overcome this problem is to document your time and money requirements for developing accurate schedules and send that information to management. If the project fails, you have documentation to support your claim that "some of these problems could have been avoided if you had given me the time and money to do effective planning." The key is not to assume total responsibility for failure, especially when the cause should be shared with others.

Closely related to the last point, inadequate resources can make it difficult for project managers to develop meaningful schedules. This circumstance is often indicative of little or inadequate management support. Projects having management support have sufficient resources necessary to conduct scheduling activities.

If you lack adequate resources, you have two options. One is to document your needs in a memo and submit it to your chain of command. Note your exact requirements and list the consequences if management fails to meet them. The second option is to do the scheduling yourself. While this mode of operation will increase your workload, it does have the advantage of reducing manpower and equipment costs.

Lack of knowledge about the goals, objectives, and requirements of the project is another reason that good schedules are not developed. Project managers facing this handicap do not have the information to do an effective job. They should identify all the shortcomings in a memo or a report to the people who can help you. As a project manager, you should list specifically your requirements and why you need this information to do scheduling. Avoid shyness over submitting your requests. Strive for as much visibility as possible by circulating the report to people who have the clout to solve your problem.

Finally, some project managers do not develop good schedules for fear of committing themselves. If they make their schedules visible, they feel they would commit themselves to meeting certain dates. And if they were to miss those dates, they would appear unreliable.

Such resistance is hard to overcome. While avoidance may prove politically advantageous, it does little good for the organization. Project managers who behave this way typically are reactive. Their projects are often plagued with sliding schedules and budget overruns.

About the only way to overcome this circumstance is for management to make scheduling mandatory. It should be standard operating procedure for projects to develop useful schedules prior to the start of any significant work. Performance appraisal of schedules should also be instituted.

Network Diagrams

Two types of schedules exist: bar charts and network diagrams. Network diagrams have their origins back into the late 1950s. They are known widely as PERT (program evaluation and review technique) and CPM (critical path method).

PERT had its origins in the Polaris missile program. It placed emphasis on fulfilling major milestones rather than following precise time estimates. CPM had its origins in the construction of chemical plants by DuPont. It placed emphasis on meeting precise time estimates.

Over the years, these two methods have merged and now have evolved into two network scheduling techniques. These two techniques are the arrow diagramming method and the precedence diagramming method.

Arrow Diagramming Method (ADM)

This method is the traditional one. It has been used widely since its origin in the 1950s. The construction industry uses this method regularly.

The arrow diagramming method is shown in **Exhibit 6.4-1**. It illustrates information using both symbols and narrative descriptions. The diagram contains a multitude of circles, or nodes, and arrows. Those symbols have important meanings. The node represents an event, such as the start or completion of an activity or a group of activities. An event consumes no time or resources. Each node contains a unique number that represents it. An arrow connects two nodes to represent an activity, such as activity Perform A. An activity consumes time and resources. Above the arrow is a description of the activity. Directly beneath the arrow is a number indicating the duration of the activity, typically in days although it can be in another increment. The numbers at the end of each arrow, contained within a node, uniquely represent the activity and are known as the I-J number. Looking at **Exhibit 6.4-1**, one can see that Perform B has an I-J number of 2-3.

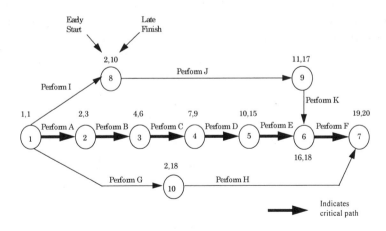

ARROW DIAGRAM FORMAT
EXHIBIT 6.4-1

Every activity within a network diagram has four dates. These are the Early Start (ES), Early Finish (EF), Late Start (LS), and Late Finish (LF). The ES is the earliest time an activity can begin; the EF is the earliest finish an activity can finish; the LS is the latest time an activity can start; and the LF is the latest time an activity can finish. These times are discussed in the section **Calculating Dates** (p. 117).

Look at **Exhibit 6.4-1**, paying attention to Activity 2-3, Perform B. Notice the leftmost pair of numbers (2,3) above node #2. The 2 in parentheses represents the Early Start date and the 3 represents the Late Finish date. Then look at the leftmost pair of numbers (4,6) above node #3. The 3 in parentheses represents the Late Start date and the 4 represents the Late Finish date for Perform C. You would designate the early and late dates on the network diagram for each activity in the manner shown in **Exhibit 6.4-1**.

Some arrows in **Exhibit 6.4-1** are boldface. This method is commonly used to represent critical path activities. These are activities that cannot slide at any time or you will miss the project completion date. Later in the chapter, you will learn how to determine the critical path.

Often, you may find in an arrow diagram a pair of bubbles with a dotted arrow between them. This symbology represents a dummy activity. This activity is employed for occurrences where no resources are used on a project and has a duration of zero. A common example of a dummy activity is one following the laying of concrete, where the concrete is allowed to harden for a period of time. This activity would have its own I-J number and the arrow would be dotted. Sometimes, you may have more than one activity beginning or ending in a project, like the one shown in **Exhibit 6.4-2**. You can create a dummy activity titled Start, where two or more activities follow the activity. You can also create a dummy activity titled Finish, where two or more activities flow into it. Either way, the Start and Finish dummies would have zero duration.

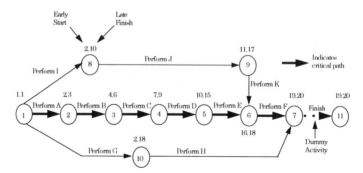

ARROW DIAGRAMMING METHOD USING DUMMY ACTIVITIES
EXHIBIT 6.4-2

Precedence Diagramming Method (PDM)

This method is an alternative to doing arrow diagrams. It is popular for projects existing in industries other than construction, such as information systems, engineering and aerospace.

Like the ADM, the PDM uses several symbols to communicate information (see **Exhibit 6.4-3**).

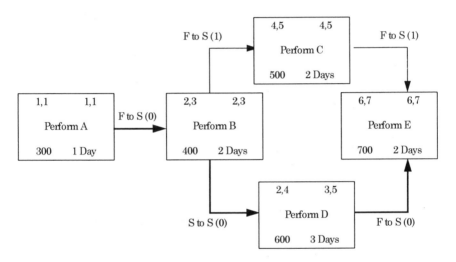

PRECEDENCE DIAGRAM
EXHIBIT 6.4-3

The first symbol is a rectangular box. This symbol represents an activity. The size of the box does not matter, though sometimes it is shown enlarged or stretched to show its impact relative to the magnitude to other activities in the project.

Within each box is a multitude of information. This information provides you with the detailed information needed to manage each activity and the entire project.

In the upper left corner of the box are the Early Start and Early Finish dates. In the upper right corner of the box are the Late Start and Late Finish dates. In the lower left corner of the box is a number uniquely identifying the activity. This number can either be the WBS code or an activity number, or identifier, that you arbitrarily assign to uniquely identify the activity. In the lower right corner is a number reflecting the duration of the activity and the increment used. In the center of the box is a verbal description of the activity, typically consisting of a command verb and object.

Symbols and information are displayed between activities. **Exhibit 6.4-3** shows an arrow flowing between activities Perform A and Perform B. This vector represents the dependency relationship between the two activities. Hence, Perform B is dependent upon Perform A (Perform B cannot begin until Perform A is complete). Perform A is known as the **predecessor** and Perform B is called the **successor**. Some activities can be both a successor via their relationship to one activity and a predecessor to another. **Exhibit 6.4-3** shows that Perform B, while a successor of Perform A, is a predecessor for Perform C and Perform D. (Perform C cannot begin until Perform B is complete, and Perform D cannot start until activity Perform B commences.)

Each arrow contains information reflecting the different types of relationships between two activities. These relationships take one of three forms: F to S; S to S; and F to F. F to S means the relationship between two activities is finish-to-start. For example, Perform B cannot begin until Perform A is 100 percent complete. S to S means the relationship between two activities is start-to-start. For example, Perform D cannot begin until Perform B has started but not completed. F to F means the relationship between two activities is finish-to-finish. Two activities must finish at the same time (not shown in **Exhibit 6.4-3**).

The number in parentheses shown next to the relationship type is known as lag. It represents the amount of "dead time" between the finish date of the predecessor activity and the start date of the successor. **Exhibit 6.4-3** shows the relationship between Perform B and Perform C as being F to S, but it has a 1 in the parentheses. Assume the duration is in days. Perform C cannot begin until 1 day after Perform B is complete. If the number was 0, Perform C could begin immediately after Perform B is complete.

Exhibit 6.4-3 shows the relationship between Perform B and Perform D as S to S and the number in parentheses is 0. When Perform B begins, Perform D can begin the same day but Perform B must start first. When reviewing a PDM diagram, always think of predecessor relationships rather than successor relationships. For instance, think of Perform A as the predecessor of Perform B, Perform B as the predecessor of Perform C and Perform D, Perform C as the predecessor of Perform E, and Perform D as the predecessor of Perform E. This way of thinking lessens the chance of becoming confused when interpreting the diagram.

As in the ADM, certain arrows are boldface to represent the critical activities of the project. You cannot slide any of these activities; otherwise, you will not meet the project completion date. This chapter discusses later how to determine the critical path.

Calculating Dates

Every activity, whether using ADM or PDM, has four dates.

1.   **ES (Early Start):**   earliest time an activity can start
2.   **EF (Early Finish):**   earliest time an activity can finish
3.   **LS (Late Start):**   latest time an activity can start
4.   **LF (Late Finish):**   latest time an activity can finish

We will use PDM to illustrate how to calculate these dates (the same method is used for the ADM). In **Exhibit 6.4-4**, you have a project involving four activities: Perform A; Perform B; Perform C; and Perform D. Assume the project begins on day 1 rather than day 0. (You could use the latter number and the process would remain the same.) Also assume that all activities have an F to S relationship with a lag value of 0. Keeping those assumptions in mind, you can begin calculating the early dates for the activities in the network diagram. Always calculate the early dates first. This process is known as the **forward pass** and involves going from left to right through the network diagram. It does not matter which path you follow through the network diagram; it does matter, however, that you go through all paths.

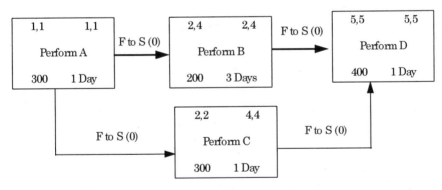

Activity Number	Activity Description	Duration Days	Early Start	Early Finish	Late Start	Late Finish	Float	Critical Path?
100	Perform A	1	1	1	1	1	0	Yes
200	Perform B	3	2	4	2	4	0	Yes
300	Perform C	1	2	2	4	4	2	No
400	Perform D	1	5	5	5	5	0	Yes

CALCULATING SCHEDULE DATES AND FLOATS
EXHIBIT 6.4-4

Perform A has a duration of 1 day, meaning the activity will begin on day 1, take the whole day, and end on day 1 for an ES of 1 and an EF of 1. Perform B is the next activity along the top path and has a duration of three days. Since Perform A finished on day 1, Perform B will begin on day 2, giving it an ES of 2. It also has a duration of 3 days. Therefore, the EF for Perform B is day 4 (because it begins on day 2, occurs also on day 3, and ends on day 4, for a duration of 3 days).

Now, take the other path. You know that Perform A has an EF of 1. Since the relationship between that activity and Perform C is F to S with a lag value of 0, Perform C begins the next day, which is day 2. Perform C has a duration of 1 day. Therefore, Perform C begins on day 2 and finishes on the same day, for a total of 1 day. The ES is 2 and the EF is the same. Notice that Perform B and Perform C are the predecessors for Perform D, meaning that Perform D cannot begin until Perform B and Perform C are complete. Whenever you confront this situation, look at the predecessor of the activity with the furthest EF date. In this case Perform B has an EF date of 4, meaning that Perform D begins the next day, giving it an ES of 5. Perform D has a duration of 1 day, thereby beginning on day 1 and ending on the same day, for an EF of 5.

With the forward pass complete, you can begin the backward pass to calculate the LS and LF dates. This process involves proceeding from left to right through the network diagram. You begin with the last activity in the network diagram. The last activity in the network diagram always assumes the LS and LF dates are the same as the ES and EF dates.

This activity, by virtue of its position in the network diagram, will always be on the critical path. Performing the backward pass is much more difficult than performing the forward pass because it requires "thinking in reverse"— a form of logic that takes considerable practice to master. Looking at **Exhibit 6.4-4**, you can see that Perform D has an LF of 5 and an LS of 5. Going through the top path first (you could have just as easily gone through the bottom path), you know that if Perform D has a LS of day 5, Perform B must have an immediate LF of 4. Perform B has a duration of 3 days, meaning that it has an LS of 2 (day 4, then day 3, and finally day 2).

Now calculate the LF and LS dates for Perform C on the bottom path. If Perform D has an LS of 5, Perform C ends up with an LF date of 4, a duration of 1 day and an LS of 4. Perform B and Perform C are the successors for Perform A (both activities depend on Perform A). With this knowledge, you can see Perform B has an LS of 2 and Perform C has an LS of 4; you need to consider the furthest LS date to the left, which is Perform B, with a LS of 2. If Perform B has an LS of 2, then Perform A has an LF of 1, because it has an F to S relationship and a lag value of 0. Since Perform A has a duration of 1 day, it has an LS of 1.

When calculating the early and late dates through a network diagram, keep the following concepts in mind.

1. The ES date is the immediate first day of an activity.
2. The EF date is calculated by taking the ES date, adding the duration, and subtracting a value of 1. Hence, EF = (ES date + Duration) – 1.
3. The LS date is calculated by taking the LF date, subtracting the duration, and adding a value of 1. Hence, LS = (LF date – Duration) + 1.
4. The LF date of an activity is the furthest time that precedes the LS date of the successor activity. Under some circumstances, you will face constraint dates. These dates are imposed upon specific activities, for instance, by mandating that an activity must start or end on a given date. Or the constraint date may require that an activity start or finish within a range, such as sometime before or after April 15th.
5. Six kinds of constraint dates exist. They are: start no earlier than, start no later than, finish no earlier than, finish no later than, mandatory start, and mandatory finish. As a project manager, you must manipulate your schedule to help you develop a network diagram that accounts for these constraint dates. This can be done by: changing the logic relationships between activities, changing activity duration, and other measures. You must remember to incorporate these constraint dates if you expect to have a realistic schedule.

Float

The early and late dates give you not only the dates an activity can start and finish; they also give the amount of time an activity can be allowed to slide past these dates before the delay becomes serious, in other words, the time you can allow an activity to slide without fear of impacting subsequent activities and, ultimately, the project completion date.

Float for an activity is calculated one of two ways. You can subtract the ES date from the LS date (LS – ES) or subtract the EF date from the LF date (LF – EF). Either way yields the same amount of float. It is customary to use the first formula.

Referring back to **Exhibit 6.4-4**, you will notice the floats for each activity have been calculated. This was done by subtracting the ES date from the LS date. Some activities have a float of 0, specifically Perform A, Perform B, and Perform D. These activities are known as your critical path.

Critical Path

The critical path illustrates those activities in the network diagram that cannot slide at any moment in time. You must complete them according to the schedule dates you have calculated.

Critical path items have some unique characteristics that distinguish them from other activities. Critical path items are located on the longest path in the network diagram. Referring to Exhibit 6.4-4, notice that the sum of the durations in the top path (Perform A, Perform B, and Perform D: 1 day + 3 days + 1 day = 5 days) is greater than the bottom path (1 day + 1 day + 1 day = 3 days), which includes Perform A, Perform C, and Perform D. You will also observe that the ES and LS dates are the same for each activity and likewise for the EF and LF dates. For example, Perform B has an ES of 2 and an LS of 2 as well as an EF of 4 and an LF of 4. Finally, all items on the critical path have the lowest float. Typically, all critical path activities have 0 float, meaning that these activities cannot miss any start or completion dates without impacting subsequent activities and the project end date. Referring to Exhibit 6.4-4, you can see that activities Perform A, Perform B, and Perform D each have 0 float, or slack.

Keep these following ideas in mind when considering the critical path:

1. Always give priority to activities on the critical path. Address those activities first, for example, by giving them your total attention and allocating resources.
2. The critical path in a network diagram constantly changes throughout a project. These changes occur because some activities will start and finish either according to the schedule or at variance with it.
3. The lower the float, the more critical the activity. On some projects, some critical path items will have a negative float. The greater the negative float, the more critical the activity.
4. A network diagram can have more than one critical path. What is important for determining the critical path is not the **total** number of activities but the **cumulative** sum of the durations of activities on any given path.

How to Construct Network Diagrams

Developing network diagrams is not easy; it requires planning. The experience can go smoothly or easily turn into a disaster.

The work breakdown structure (WBS) is your primary source of information for the network diagram. You do not, however, use all the items listed in the WBS; you use only those activities located at the lowest levels on each leg.

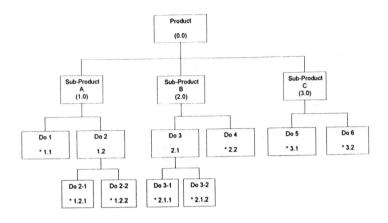

WORK BREAKDOWN STRUCTURE INDICATING WORK PACKAGE LEVELS
EXHIBIT 6.4-5

These items are known as the work package level and are the common elements you used to build your network diagram. **Exhibit 6.4-5** shows the work package levels for a WBS chart.

After identifying the work package level elements, you begin the laborious task of connecting them to one another to reflect a logical sequence. A common method

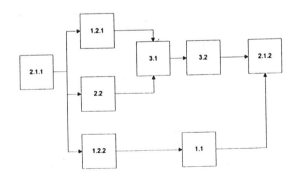

NETWORK DIAGRAM USING WORK PACKAGE LEVEL ITEMS
EXHIBIT 6.4-6

is to write only the activity number or WBS code in the box representing an activity. An example of this logical sequence is shown in **Exhibit 6.4-6**.

There are two common ways to develop the network diagram. The first way is to sketch the diagram on a large sheet of white paper (taping several easel pad–size sheets of paper will do) using a pencil and eraser. You continue drafting the logical sequence until you have it correct.

Another way is to use a large sheet of paper (taping several easel pad–size sheets of paper will do) on a large wall. Then, you record the activity number or WBS code for each task on a separate tag (a Post-it is fine). You place the applicable tag on the paper in a position that seems appropriate in relation to the other ones. Once your are sure the logical sequence is correct, draw the arrows connecting each one of the WBS elements. The result is a draft of your first network diagram.

---

[1] The numeric designator uniquely identifying the activity

[2] A short narrative description of the activity

[3] A constraining date which may be one of these:
- start no earlier than
- start no later than
- finish no later than
- finish no earlier than
- mandatory start
- mandatory finish

[4] The constraint date in [3]

[5] The activity number of the activity preceding the activity shown in column [1]

[6] The type of relationship between the activity listed in column [1] and its predecessor shown in column [6]. The relationship is either Finish-to-Start (F to S), Start-to-Start (S to S), or Finish-to-Finish (F to F)

[7] The number of days of "dead time" between the finish date of the activity in column [5] and the start date of the activity shown in column [1]

[8] The activity number of the activity following the activity shown in column [1]

[9] The type of relationship between the activity listed in column [1] and the successor shown in column [9]. The relationship is either F to S, S to S, or F to F.

INSTRUCTIONS FOR COMPLETING THE ACTIVITY DESCRIPTION FORM
SHOWN IN EXHIBIT 6.4-7

Activity Number	Activity Description	Imposed Date Type	Imposed Date	Predecessor(s)	Relationship Type	Lag	Successor(s)	Relationship Type
[1]	[2]	[3]	[4]	[5]	[6]	[7]	[8]	[9]
2.1.3	Determine customer needs	Mandatory start	95.03.18	2.1.2	FS	0	2.1.4	FS
2.1.4	Review with team	—	—	2.1.3	FS	0	2.2.1	SS

ACTIVITY DESCRIPTION FORM
EXHIBIT 6.4-7

Regardless of your method of constructing the network diagram, you should have it recorded, graphically, on paper (preferably no larger than 11 × 17 inch paper). You can then distribute copies of the diagram to everyone who has an interest.

If you are using an automated scheduling package, then you can print excellent network diagrams using a printer or plotter. Remember, however, that you still have to construct the WBS, make time estimates, determine the logical sequence of activities, and enter all that information into the computing system. No software package exists that will do all these activities for you.

You can use the form in **Exhibit 6.4-7** to help you develop your network diagram. Just match the applicable numbers identified in the exhibit.

Bar Charts

Bar charts, also called **Gantt charts**, are more widely used in many different industries. These charts are easy to construct and convey information more clearly and concisely. An example of a bar chart is shown in **Exhibit 6.4-8**. These charts do not provide all the information communicated in a network diagram. Too often, some project managers overlook these shortcomings.

Bar charts do not show the dependencies among activities. You cannot tell which activity is the predecessor to another activity or activities. Nor can you tell which activities are the successors. Bar charts do not show all four dates for each activity. Normally, the beginning of the bar shows the early start date and the end of the bar shows the early finish date. Rarely do project managers construct bar charts with bars reflecting the late start and late finish dates for activities.

Another shortcoming exists: bar charts do not indicate which activities are

Task	Re-spon-sibil-ity	Year	19XX		
		Month	April	May	June
		Week	1 2 3 4	1 2 3 4	1 2 3 4
Perform A	RLK		▬▬▬ 100%		
Perform B	ISL		▬ 25%		
Perform C	RLK		▭		
Perform D	KLR			▭	
Perform E	KLR			▭	
Perform F	ISL				▭
Perform G	RLK				▭

TYPICAL GANTT BAR CHART
EXHIBIT 6.4-8

critical and which are not. The critical path is not clearly shown unless symbols are used to designate which activities are critical and which are not. In addition, bar charts do not show the float for each activity. This shortcoming makes it difficult to determine which activities require immediate attention.

Despite these shortcomings, bar charts do have their uses. They are excellent for reporting to senior management and the client. They are easier to read and understand than network diagrams. And they are perfect for providing summary schedule information to both management and the client. Bar charts enable you to summarize different levels within the work breakdown structure, whereas network diagrams entail only the lowest-level items in the work breakdown structure, known as the work package level.

Any bar charts you use should contain certain features common to all bar charts; these are reflected in **Exhibit 6.4-9**. First, the chart must have a time scale to show the flow-time, or duration, of each activity. This time scale can be in any increment, such as hourly, daily, weekly, quarterly, semi-annually, or yearly. Second, the bar chart must have a column to list the tasks. This column is subdivided into two subcolumns. One is for listing the activity number that uniquely identifies the activity and the other for containing a narrative description of the activity. Third, bars should be drawn to reflect the flow-time, or duration, of each activity. You can use solid or hollow bars, or both. Hollow bars can be used to represent the entire activity, and the percentage of completion for a task can be represented by shading the bar accordingly.

Many approaches have been used to modify bar charts to overcome their shortcomings. **Exhibit 6.4-9** shows some of these features.

					Year		19XX									
					Month	April				May				June		
					Week	1	2	3	4	1	2	3	4	1	2	3	4
Task No.	Task Descrip	Total Hours	Float Hours	Predecessor(s)													
100	Perform A	80	0	None				100%									
200	Perform B	20	1	100					25%								
300	Perform C	15	5	200													
400	Perform D	10	0	100													
500	Perform E	90	0	400													
600	Perform F	20	10	300,500													
700	Perform G	20	0	500													

GANTT (BAR) CHART SHOWING WAYS TO OVERCOME SOME SHORTCOMINGS
EXHIBIT 6.4-9

How to Construct a Bar Chart

Building a bar chart is easier than building a network diagram. You still use the WBS structure for your source of information, preferably at the work package level. Unlike for network diagrams, you can build bar charts showing bars for different levels within the WBS. You can show bars for elements at the same levels in the WBS or mix levels using the WBS code of task descriptions to indicate the level within the WBS, as shown in **Exhibit 6.4-10**.

WBS	Task	Year	19XX		
		Month	April	May	June
		Week	1 2 3 4	1 2 3 4	1 2 3 4
0.0	Product				
1.0	Sub-Product A				
1.1	Do 1				
1.2	Do 2				
1.2.1	Do 2-1				
1.2.2	Do 2-2				
2.1	Sub-Product B				

BAR CHART SHOWING HIERARCHY OF ELEMENTS WITHIN THE WBS
EXHIBIT 6.4-10

You can draw most bar charts on a large sheet of paper (such as 11″ × 17″), remembering to include a column for task identification and an area for drawing the bars to reflect the flow time for an activity.

When drawing the network diagram, pay attention to the logical sequence of the tasks. It is a common mistake to draw bars concurrently when in fact they should follow each other sequentially. Project managers sometimes draw a bar for a task, not clearly understanding the time sequence for the activities. Releasing a schedule with that fundamental error can cause some embarrassment.

You have several options available to improve the appearance of your draft bar chart. Some computer software packages allow you to enter the dates for an activity. Others let you enter the information on larger systems using graphic packages, giving the bar charts a professional appearance. Regardless of appearance, most important is the accuracy of the information, not the pretty appearance of the schedules.

Developing a Schedule

Whether you develop a bar chart or a network diagram, remember the following points. Never put a draft schedule in final form. It psychologically impacts the quality of feedback you receive from people. People are reluctant to criticize something in final form. They feel you have done considerable work—and you may have—to build the schedule. If they see something incorrect, they may feel that changing it may require considerable effort and may hurt your feelings. Keep the schedule in draft form until it has received complete approval.

Utilize both individual sessions and group meetings to build the schedule. Nothing says you must build the schedule yourself. Indeed, building a schedule independently can exposure you to the objections of others later in the project. The idea is to create a schedule that everyone can agree to follow. The best approach is to determine who should participate in building the schedule. Ideally, you will have identified all those individuals after creating the WBS. You should sit with each person, individually at first, and draft the portion of the schedule pertaining to them. Having received feedback from everyone, you can hold a group meeting(s) to resolve disagreements. After achieving consensus, obtain everyone's signature on the schedule. Record the date of signing, too. This action signifies agreement to work to the schedule. Obtaining signatures is more than receiving their approval; it also has a beneficial psychological impact. When people sign a document, they commit themselves in a form that is recorded.

Armed with signatures from everyone, you produce the schedule in final form. You should store the original in a safe place and give copies to each of the members on the project and other interested parties. The worst action you can take is to develop a schedule and hide it from the team. Make sure everyone who needs a schedule has one. Just because you have the schedule in final form does not mean you cannot change it periodically. Occasionally, you may have to reschedule. You should archive previous versions of the schedule when rescheduling. This action provides excellent traceability of what has caused rescheduling to occur. It also provides invaluable archival information for future projects of a similar nature.

Conclusion

Whether a bar chart or network diagram, your schedule is your road map for completing your project. Any inaccuracies in it can lead you astray. You must expend considerable effort, time, and money to develop a valid schedule. Otherwise, you are likely to find yourself heading down the road leading to failure rather than success.

30400 SCHEDULES

Political Pressures	Work Breakdown Structure	Time Estimates	Statement of Work	Risk Identification	Major Participants	Allocated Resources	P² M² CYCLE — TASKS	Project Manager	Senior Management	Project Sponsor	Project Team	Client	Procedures	Network Diagram	Bar Chart
					■		**30405** Develop procedures for performing scheduling	■					■		
			■				**30410** Determine type of schedule	■						■	■
					■		**30415** Determine whether to use individual or group sessions or both to develop the schedule	■					■		
■							**30420** Determine the pressures you face that could prevent you from developing a realistic schedule and dealing with those pressures	■						■	■
	■				■		**30425** Determine the method for developing the schedule	■					■		
	■						**30430** Apply work breakdown structure	■			■	■		■	■
	■	■	■			■	**30435** Determine logical sequence	■			■	■		■	■
		■	■				**30440** Apply estimates	■			■	■		■	■
	■	■	■	■			**30445** Calculate dates	■			■	■		■	■
	■	■	■	■			**30450** Determine critical path	■						■	
	■	■	■	■			**30455** Calculate float	■						■	
	■	■	■	■			**30460** If developing a Gantt chart, decide whether to reflect early or late start and finish dates	■							■
					■		**30465** Determine who must review the schedule	■			■	■	■		
					■		**30470** Determine who must approve the schedule	■	■	■		■	■		
■							**30475** Apply revision control over the schedule	■					■		
					■		**30480** Determine responsibility for scheduling	■					■		
							MEASURES OF SUCCESS								
							Is the appropriate type of scheduling being developed?							■	■
							Are the estimates used to build the schedule?						■	■	■
							Is the WBS used for the schedule?						■	■	■
							Are the early start and stop dates calculated?							■	■
							Are the critical activities identified?							■	
							Have all the major participants given their concurrence to the schedule?						■	■	■
							Does the schedule provide the path(s) to achieve project goals and objectives?							■	
							Does the schedule reflect all the date constraints identified in the SOW?							■	■
							Does everyone responsible for performing tasks and achieving milestones have a copy of the schedule?						■		
							Are procedures for scheduling documented?						■		

6.5 RESOURCE ALLOCATION (30500)

Once you have your estimates and network diagram in place, you can start assigning your available resources to the tasks in the schedule.

Typically, you have four categories of resources to assign: labor, equipment, material, and supplies. Sometimes, you may find it necessary to assign the cost of facilities to one or more tasks. In this chapter, we will cover assigning labor to tasks. The same principles apply to many non-labor resources.

You have other resources at your disposal besides labor, equipment, material, supplies, and facilities. The two most common of these are time and money. Both must be closely watched throughout a project.

Thoughts on Resource Allocation

When allocating resources, regardless of type, remember to give preference to critical path activities. These are the activities most important for completing the project on time. Giving resources to activities of less significance illustrates poor judgment. Disregarding the critical path results in waste and lower productivity as resources are assigned to where they are least needed.

When facing concurrent activities on the critical path, give preference to the activity with the least float when allocating resources. Also, use the same criterion to allocate resources among activities on the path. Activities with the least float are behind schedule the most, thereby most often requiring immediate attention.

Give preference to the most complex activity when the float is equal for concurrent activities, whether you are comparing activities on a critical or even a path. More complex activities tend to consume larger amounts of resources than easier activities.

The Resource Histogram

After assigning resources, specifically labor, to tasks in the schedule, you can develop a **resource histogram**, as shown in **Exhibit 6.5-1**.

The resource histogram profiles how well you would use your staff given the current schedule. Frequently, the histogram has an irregular pattern, which can later prove significant when you are attempting to smooth it.

Constructing a histogram can be done manually or through the use of project management or graphics software. Project management software is the more useful way to produce the histogram because it is linked to your WBS. If you need to draw a histogram manually, just follow these three simple (but somewhat time-consuming) steps.

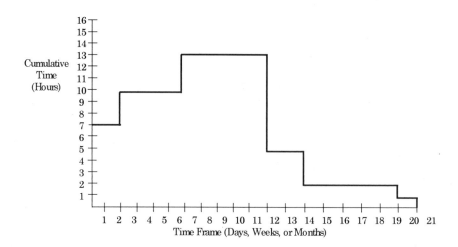

RESOURCE HISTOGRAM – UNLEVELED
EXHIBIT 6.5-1

1. Draw an X-axis, or horizontal line, across the page. This line represents a time-scale in increments of your choosing.
2. Draw a Y-axis, or vertical line, perpendicular to the X-axis and at its left end point. This line represents the cumulative time, usually in hours, that you will use certain resource(s) at a specific point in time. Where the X- and Y-axes meet is called the point of origin and represents 0 in terms of time and effort.
3. Draw a horizontal bar to represent the cumulative time a resource will work during each time unit on the X-axis.

How to Interpret a Histogram

With your histogram, you decide whether you like it. If the profile has an irregular shape, such as the one in **Exhibit 6.5-1,** but you still like it, then you need not modify either your schedule or resource allocations. If you decide not to change anything, consider the potential problems that often accompany projects with histograms that look like the one in **Exhibit 6.5-1**.

The peaks in the histogram may result in serious problems. They indicate that the existing staff will need to work overtime at certain periods in the project. This circumstance can quickly lead to higher labor costs and burnout.

The valleys show the opposite picture. Employees may not be fully productive and may even be sitting idle for long periods. This circumstance can also lead to problems. Idleness of employees often translates into lower morale. Some of these inactive team members may interfere with productive employees, or the productive employees may feel they are assuming an unfairly large share of the workload.

The irregular pattern also represents other dysfunctional circumstances. Project managers must continually acquire and release team members, thereby preventing them from concentrating on more important matters. If they do not acquire and release continually under such circumstances, they are forced to absorb needless labor costs (unless they work in a matrix environment).

The irregular pattern also reflects another potential problem. Project managers may become too dependent on staff members with rare, specialized skills. If these specialists quit at a peak period, their departure could negatively impact the schedule. To lessen this dependency, project managers need to acquire more people with specialized skills, cross-train people, and lower the peak (known as resource leveling).

Accepting a resource histogram with the pattern shown in **Exhibit 6.5-1** does not necessarily mean disaster. You can take positive action to counter problems like the ones mentioned above. For peak labor usage, you could decide to use outside consultants to perform a portion of the workload, thus alleviating stress on employees. You could also elect to concentrate only on critical activities and let the ones slide for a period of time. For valleys, you could send people to training; encourage them to take vacations, or use them to support other people working on critical activities. The point is that the irregular pattern in **Exhibit 6.5-1** is fine if you can manage it. But if you fail to do so, you may end up with a management style that has all the symptoms of management by crisis and management by drives.

Resource Leveling

Resource leveling allows you to optimize the use of your resources and to avoid having a histogram riddled with peaks and valleys. Through resource leveling, you have a histogram that is smooth, or level, like the one shown in Exhibit 6.5-2.

A **leveled histogram** allows you to use resources more efficiently and effectively. Through leveling, you remove the steep peaks and valleys in the histogram. You will not have people sitting idle as often as you would if you had kept an unleveled histogram. And you will not have people working overtime as often either, thereby reducing labor costs and personnel burnout.

The level histogram allows the project manager to concentrate on production rather than on continually worrying about acquiring and releasing re sources. You will have a steady work force to support you and will not need to supervise the project team as closely. You will be able to delegate work more effectively

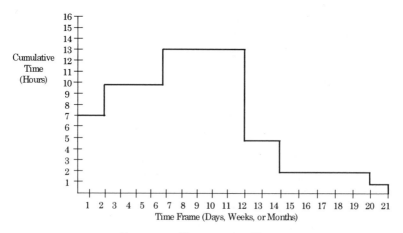

RESOURCE HISTOGRAM - UNLEVELED
EXHIBIT 6.5-1

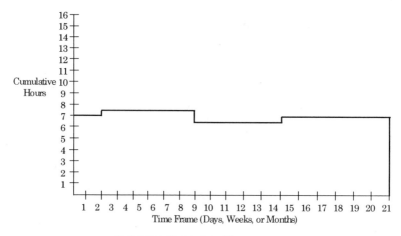

LEVELED RESOURCE HISTOGRAM
EXHIBIT 6.5-2

and focus your attention on the important aspects of the project. Resources represented by a level histogram permit you and your work force to become more productive.

The level histogram also allows you to function proactively rather than reactively by minimizing the opportunity for management by drives and management by crisis.

Ways to Level a Histogram

You have several options available to level your histogram.

One way is to change the logical relationships among activities. For instance, you might change the relationship between two activities from start-to-start to finish-to-start, as shown in **Exhibit 6.5-3**.

You can also change the dependencies between activities. For instance, you might make an activity dependent upon a different activity to lower the peak in the histogram, as shown in **Exhibit 6.5-4**.

Another way to level your histogram is to lengthen the lag value between two activities. This option will increase the spread from the time an activity finishes and when the subsequent one (successor) can begin, but it can effectively lower a peak in the histogram. However, it may also extend the project completion date (which is often unacceptable). **Exhibit 6.5-5** illustrates this option.

Still another option is to extend the duration (flow-time) of a activity while reducing proportionately the hours per day to work on it. For example, an activity originally estimated to take 80 hours of work over 10 days at 8 hours per day may be changed to 20 days at 4 hours per day. Regardless, the cumulative hours to complete the task remains the same, 80 hours. This option rarely applies to activities on the critical path. If you apply it to critical path tasks, the project completion date will slide.

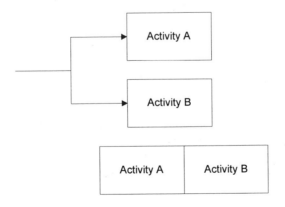

TWO ACTIVITIES WITH START-TO-START RELATIONSHIP
CHANGED TO FINISH-TO-START TO LEVEL HISTOGRAM
EXHIBIT 6.5-3

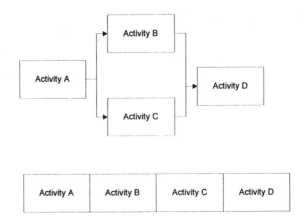

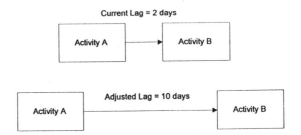

CHANGING THE LOGIC RELATIONSHIP BETWEEN ACTIVITIES
IN A NETWORK DIAGRAM TO LEVEL A HISTOGRAM
EXHIBIT 6.5-4

CHANGING THE LAG VALUE BETWEEN TWO ACTIVITIES TO LEVEL A HISTOGRAM
EXHIBIT 6.5-5

An effective but lazy option to level your histogram is to eliminate activities from the schedule; you simply remove their presence. This option will lower the peaks in the histogram but could result in problems later in the project. Removing an activity because it complicates your situation may momentarily prove advantageous. Later, however, you may find that not performing the task has negative consequences.

You can also reduce the hours assigned to perform a task and compensate for that reduction by purchasing technology. This option works only if the team members are already skilled in the new technology. If they are not proficient, then the time needed to complete the task may actually increase.

Another approach is simply to reduce the number of hours per work day and the cumulative total without adding any automated equipment to a task. This is based on the premise that people will work harder and smarter to complete the task and is based on Parkinson's Law—that a task will fill the time available for its completion. Hence, if you authorize 100 hours to do Activity A, it will take 100 hours to complete. If you authorize 75 hours to complete the same activity, your people will finish the task in 75 hours.

Parkinson's Law works only to a certain point. Some tasks simply cannot be completed in a shorter time. If good historical information exists (providing a reliable estimate), you may find it unwise to lessen the estimate. A smaller estimate could mislead everyone on the project, resulting in needless pressure and a substandard product.

Resource leveling is easier if the project completion date is not mandated (i.e., the project need not finish on or by a specific date). You can exercise any of the options discussed above until you have leveled your schedule. However, if the project completion date is fixed, leveling your schedule is more complicated. In fact, you may have to accept some high peaks and low valleys in your histogram.

You can use the form shown in **Exhibit 6.5-6** to record the information necessary for building a resource histogram or using labor. Just complete the instructions on the form.

Conclusion

A resource histogram is a tool you can use to maximize resource utilization, whether for labor, supplies, equipment, etc. For labor, you can generate a histogram for each person, by occupation, or for the entire team. You can then level the histogram to optimize the utilization of one or several persons cost effectively.

---

[1] The last name, first name, and middle initial of the person assigned to the activity
[2] The position or title of the person to work on the activity, such as a system analyst or laborer
[3] The average number of hours each day that you expect the person to work on the activity
[4] The unique designator identifying the activity in question
[5] The number of working days for the activity
[6] The earliest time that the activity can begin
[7] The earliest time that the activity can complete
[8] The latest time that the activity can begin
[9] The latest time that the activity can complete

INSTRUCTIONS FOR COMPLETING RESOURCE HISTOGRAM FORM SHOWN IN
EXHIBIT 6.5-6

30500 RESOURCES ALLOCATION

P² M² CYCLE

Resources	Work Breakdown Structure	Time Estimates	Statement of Work	Risk Identification	Major Participants	Network Diagram	Bar Chart	TASKS	Project Manager	Senior Management	Project Sponsor	Project Team	Client	Resource Analysis	Procedures	Revised Schedule	Resource Histograms
					■			**30505** Establish a procedure for resource acquisition and allocation	■						■		
■								**30510** Determine the categories of resources	■					■			
■								**30515** Determine the unit cost for each resource category	■					■			
■			■					**30520** Determine the source of each resource category	■					■			
■								**30525** Establish the required quantity of each resource category	■			■	■	■			
■								**30530** Estimate the required quantity of each resource category per task	■			■	■	■			
■	■	■				■	■	**30535** Generate histograms for each category of resources	■								■
■	■	■				■	■	**30540** Generate histograms for each specific resource	■								■
■	■	■		■		■	■	**30545** Interpret resource histograms	■			■	■	■			■
					■			**30550** Assign responsibility for resource acquisition and tracking	■			■	■		■		
								30555 Determine when to acquire resources with adequate lead time	■			■	■	■			■
■						■		**30560** Give preference to critical path	■					■			■
■						■	■	**30565** For concurrent activities, give preference to the activity with the least float	■					■			■
■						■	■	**30570** For concurrent activities with the same float, give preference to the most complex activity	■					■			■
■						■	■	**30575** Perform leveling, where necessary	■							■	■
■								**30580** Determine the resources that you will need	■			■	■	■			
■								**30585** Determine the resources that you have available	■			■	■	■			
■								**30590** Identify potential contractors	■					■			
■								**30595** Verify credentials of contractors	■					■			
					■			**30599** Determine responsibilities for resource allocation	■						■		
								MEASURES OF SUCCESS									
								Does the resource allocation activity provide reasonable confidence in achieving project goals and objectives?						■			■
								Does the resource allocation activity reflect constraints identified in the SOW?						■		■	
								Is the schedule used for resource allocation?						■		■	
								Do the histograms reflect an attempt to allocate resources in a level manner?						■			■
								Have all the major participants given their concurrence to the allocation of resources?							■		
								Is the resource allocation activity done for the entire project as well as for individuals?						■			

Resource Description	Resource Type	Hours / Day	Activity Number	Duration	Early Start Date	Early Finish Date	Late Start Date	Late Finish Date
[1]	[2]	[3]	[4]	[5]	[6]	[7]	[8]	[9]
Joel Smith	Labor	6	2.1.3	1.5 days	xx/xx/xx	xx/xx/xx	xx/xx/xx	xx/xx/xx
Lorna McNeil	Labor	8	2.1.4	2 days	xx/xx/xx	xx/xx/xx	xx/xx/xx	xx/xx/xx

RESOURCE HISTOGRAM RESOURCE FORM
EXHIBIT 6.5-6

6.6 COST CALCULATION (30600)

Even with your statement of work, work breakdown structure, network diagram, bar charts, time estimates, schedule dates, and resource histograms, you still must determine whether your plans fall within cost constraints.

Money is an important factor for any project. Indeed, without money, your project could never be a reality. It provides you with the "energy" to lead, define, plan, organize, control, and close the project. You should manage it with the same care as all other types of energy, avoiding any waste. Because there is seldom a glut of money—and then for only a short period—money must be conserved. Even in times when plenty of money exists, it pays to maximize the output for every dollar spent.

Reasons for Using Money Wisely

There are several obvious benefits to maximizing output for each dollar spent on your project. Money can be saved for leaner times, though for projects lasting less than a year, this circumstance rarely pertains. For projects lasting more than a year, the likelihood increases that funds may later be curtailed or operations scaled back. If permitted, you can use the remaining funds to finance operations in the following year.

You can increase the company's profit margin. The less money you use, the more money there is available to contribute to company profits. A huge overhead or wasteful operation reduces the money distributed to shareholders.

You can provide more funds for critical activities. For example, you can apply remaining funds to important tasks by eliminating wasteful expenditures or activities for noncritical path items. This will give you more money to work with when dealing with critical path items.

Finally, maximizing output for every dollar spent allows a product to be priced competitively. If the selling price includes high overhead costs, the asking price for the product may exceed what customers are willing to spend.

What Consumes Money

It behooves you as a project manager to control your expenditures. To do that, you must recognize what consumes money.

Labor is a major consumer of money on a project. Because you always need people to do the work, typically labor consumes the greatest share of the money.

Training also consumes money. You may want to train people to increase their job knowledge and skill level. Training can prove expensive in the short-run; however, it can lead to greater productivity in the long run.

Equipment is another potentially costly item. People need the right tools to do their job. Blue collar workers may need a forklift to lift material. Programmers may need a personal computer and special tools to develop software. Marketing professionals may need a desktop publishing program to develop impressive documentation. Regardless, the use of equipment requires money.

Traveling, too, can consume money, especially when participants are geographically dispersed. Under such circumstances, travel may be necessary to maintain good communications.

Facilities also require money. Periodically, a project may need a structure to complete one or more tasks. Perhaps the project team needs an offsite facility to avoid interruptions. Or, it may require using a facility simply because services are unavailable elsewhere. For example, engineers may need to work at a specific laboratory that provides services necessary for them to do their work; use of those facilities must be charged to the project.

A project also requires money for supplies. No project, regardless of industry, can exist without using supplies, such as paper, diskettes, and other office equipment.

Material bears its costs, too. Low-quality material may translate into inferior products. This translates into customer dissatisfaction, lost sales, and sometimes legal ramifications.

Time also consumes money. Time must be managed just as wisely as the money appropriated for your project. It follows that a close relationship exists between time and labor. The more time you use, the greater the labor costs.

Finally, you must control expenditures on a seldom appreciated resource: information. Expenditures on information can become wasteful. Often, many project managers seek information for its own sake or they seek the wrong information. They fail to realize that it is important to acquire the right information. A frantic search for just any information can mean spending unnecessary sums of money. Just the collection of information can prove expensive.

Two Categories of Costs

Essentially, you can divide project costs into two categories: direct and indirect. Direct costs are attributable to a particular work breakdown structure (WBS) item; indirect costs are not directly attributable to a specific WBS item. An example of a direct cost is equipment; an example of an indirect cost is use of a facility.

As a project manager, you should calculate four sets of cost data. These are: estimate costs, budget costs, actual costs, and estimate-at-completion costs.

Estimate costs are calculated before the project begins. They are projections of the amount of money you expect to spend on each activity and for the entire project. Budget, or baseline, costs are expenditures you "freeze," or allocate. They are costs you agree to comply with during the project. You cannot modify these costs without re-planning and instituting change management procedures. Budget costs will match your estimates. In other words, you do the estimates first and they become your budgets.

Actual costs are incurred during your project. They represent what has been spent for each WBS element and the entire project, up to a specific point in time.

Estimate-at-completion (EAC) costs represent a combination of actual costs and estimated monetary requirements to complete tasks and even the entire project. This figure depends upon whether you are under or over your budget at a specific point in time.

Calculating Estimate Costs

Estimate costs represent the sum of expenditures to perform a task and the entire project. You determine the funding required for each task and sum those figures for the entire project. The funding is based upon expenditures for labor, equipment, training, and other items. **Exhibit 6.6-1** illustrates a method for recording estimate costs.

---

[1] The numeric designator uniquely identifying the activity
[2] A short narrative description of the activity
[3] Labor estimates = the summation of rate × hours/day × duration
[4] Equipment estimates = the summation of price × number of units × duration
[5] Supplies estimates = the summation of amount × unit price
[6] Material estimates = the summation of price × number of units
[7] Training estimates = the summation of training course × number of attendees
[8] Others:
   Facilities estimates = the summation of rate × duration
   Information estimates = usually incorporated in other estimates
   Travel estimates = the summation of carrier cost + allowances + accommodations + car rentals
[9] Cumulative cost estimates for entire task or project

INSTRUCTIONS FOR COMPLETING THE
RESOURCE COST ESTIMATE FORM SHOWN IN
Exhibit 6.6-1

Activity Number	Task Description	Labor	Equipment	Supplies	Materials	Training	Other	Totals in $'s
[1]	[2]	[3]	[4]	[5]	[6]	[7]	[8]	[9]
2.1.1	Perform A	3,000	450	100	600	1,000	100	5,250
2.1.2	Perform B	2,000	850	125	400	800	175	4,350
2.1.3	Perform C	8,500	1,700	75	300	300	800	11,675
2.1.4	Perform D	10,000	300	80	1,000	1,500	300	13,180
2.2.1	Perform E	9,000	400	35	300	500	325	10,560
2.2.2	Perform F	4,500	500	29	700	600	175	6,504
	Totals:	37,000	4,200	444	900	4,700	1,875	49,119

Costs in $'s

RESOURCE COST ESTIMATE FORM

EXHIBIT 6.6-1

Estimate costs involve determining in advance the level of expenditures for labor, supplies, equipment, training, and other items for each activity and the entire project. Typical formulas for calculating cost estimates for variables that consume expenditures are:

Equipment estimates = price × number of units × duration
Facilities estimates = rate × duration
Information estimates: usually incorporated into other estimates
Labor estimates = rate × hours/day × duration
Supply estimates = amount × unit price
Material estimates = amount × unit price
Training estimates = training course cost × number of attendees
Travel estimates = carrier cost + allowances + accommodations + car rental

The formula for estimating costs for each task is:

Estimate cost per task = labor estimates + training estimates + equipment estimates + supply estimates + material estimates + facility estimates + information estimates + travel estimates + other estimates

The formula for determining the estimate cost for an entire project is:

Estimate cost for entire project = $\sum$ estimate cost per task

where $\sum$ means the summation of each task

You can use the form in **Exhibit 6.6-2** to help you calculate your estimates regarding labor costs. Just follow the instructions shown below:

[1] The numeric designation uniquely identifying the activity in question
[2] The occupational description of the worker, such as carpenter or systems analyst
[3] The source of employment of the worker, such as the subcontractor or firm
[4] The rate of pay for the labor during regular working hours
[5] The total number of normal hours to work each day
[6] The total number of overtime hours to work each day
[7] The rate of pay for the worker during overtime periods, typically 1.5 or 2.0
[8] The sum of regular and overtime hours
[9] The sum of expenditures for labor under normal and overtime conditions

INSTRUCTIONS FOR ESTIMATED LABOR USAGE FORM SHOWN IN
EXHIBIT 6.6-2

You can also use the form in **Exhibit 6.6-3** to help you calculate your estimates regarding nonlabor costs. Just follow the instructions shown below:

[1] The numerical designation uniquely identifying the activity
[2] A short, narrative description of the activity
[3] The flow-time (number of work days) for the task
[4] A narrative description of the resource
[5] The classification of the resource, such as supplies or equipment
[6] The projected unit price for the resource
[7] The projected cumulative number of units required to complete the project
[8] The projected cumulative cost for using the resource throughout the duration of the activity; calculated by multiplying column [5] by [6] by [8]

INSTRUCTIONS FOR ESTIMATED NON-LABOR USAGE FORM SHOWN IN
EXHIBIT 6.6-3

Activity Number	Labor Type	Status	Regular Rate in $'s	Estimated Regular Hours	Estimated Overtime Hours	Overtime Rate in $'s	Estimated Total Labor Hours	Estimated Total Labor Costs in $'s
[1]	[2]	[3]	[4]	[5]	[6]	[7]	[8]	[9]
2.1.1	Systems Analyst	Staff	50	100	10	75	110	5,750
2.1.2	Business Analyst	Consultant	100	50	0		50	5,000
	Totals:			150	10	$75	160	$10,750

ESTIMATED LABOR USAGE FORM
EXHIBIT 6.6-2

Activity Number	Activity Description	Duration	Resource Description	Resource Type	Estimated Unit Cost	Estimated Total Number of Units	Estimated Total Costs
[1]	[2]	[3]	[4]	[5]	[6]	[7]	[9]
4.2.6	Build Mainframe Programs	30 days	machine time	mainframe	$20 per minute	600	$12,000
	Totals:						$12,000

ESTIMATED NON-LABOR USAGE FORM

EXHIBIT 6.6-3

Calculating Budget Costs

As mentioned earlier, the budget costs are the same as the final estimate costs. Budget costs represent "frozen" figures. You cannot change these costs without replanning, and you agree to abide by them during the project.

Calculating Actual Costs

Actual costs represents the costs incurred up to a specific point in time for each task and the entire project. The idea is to spend actual monies (known as actuals-to-date) equivalent to what you had allocated. Money spent that exceeds the amount allocated (budget) for a task is called an overrun; the opposite is called an underrun.

You determine actuals-to-date by adding the money spent reflected in your daily or weekly labor reports and purchase orders. These documents, discussed in Chapter 7, help to record what you have spent for each activity and the entire project up to a specific date. You then match these amounts to what is called the budget-to-date.

Take the activity, Perform B, in **Exhibit 6.6-4** for an example. The amount allocated is $3,950. Up to the current date, the activity is 75 percent complete. Take the 75 percent and multiply it by the budget amount of $3,950 to get $2,962. The $2,962 is the amount you should have expended for the task once it is 75 percent complete. The actual-to-date, however, is $4,000, or $1,038 more than originally expected (calculated by subtracting the budget-to-date from the actuals-to-date). This excess amount is called an **overrun** and means you have exceeded the budget for the task and possibly the project.

Under some circumstances, the difference between budget-to-date and the actuals-to-date may be less than expected (i.e., you have spent less than planned and have additional monies available). Perform F in **Exhibit 6.6-4** is an illustration of this circumstance. You had allocated $5,804 and completed 60 percent of the task, giving you a budget-to-date of $3,482 ($5,804 × 60 percent). You actually spent $3,000, thereby giving you an underrun of $482 ($3,482 − $3,000).

Overruns and **underruns** may be symptomatic of problems confronting a project. An overrun may indicate that people work on a task much more extensively than expected, thereby increasing costs. The project team may have worked too much overtime, which can dramatically increase labor costs. Or you may just be using more of other resources than you had planned.

An overrun may indicate that your project has experienced an unusually large number of problems. These problems may concern the labor force or the development of a product. For instance, you may not have received the cooperation from people instrumental in delivering a product, such as a

Activity Number	Task Description	Estimate in $'s	Budget in $'s	Percent Complete	Budget-to-Date in $'s	Actuals-to-Date in $'s	Over/Under in $'s	Totals in $'s
2.1.1	Perform A	4,650	4,650	100%	4,650	5,000	350	5,000
2.1.2	Perform B	3,950	3,950	75%	2,962	4,000	1,038	4,988
2.1.3	Perform C	11,375	11,375	60%	6,825	12,000	5,175	16,550
2.1.4	Perform D	12,180	12,180	100%	12,180	12,180	0	12,180
2.2.1	Perform E	10,260	10,260	30%	3,078	8,000	4,922	15,182
2.2.2	Perform F	5,804	5,804	60%	3,482	3,000	(482)	5,322
	Totals:	48,219	48,219		33,177	44,180	11,003	59,222

COMPLETE PROJECT COST FORM
EXHIBIT 6.6-4

document, software, or hardware. Or you may be building a technical product and find that the feat is more complicated than originally envisioned. Finally, an overrun may reflect a bad estimate. Your estimate in the first place may have been inaccurate—too optimistic.

Underruns are often construed as positive simply because less money was spent than expected. Yet, that perception can mislead everyone. An underrun in costs for resources may be as sign that the resources are sacrificing quality (i.e., that people are cutting corners to complete their work, which can have negative consequences). Or an underrun may mean the estimate was initially inaccurate— too pessimistic.

Unlike overruns, not all underruns are negative. Underruns may truly reflect actual conditions., which can very well be positive. A cost variance (the difference between planned and actual) need not always cause alarm and need not necessitate investigating why the difference exists.

Calculating Estimate-at-Completion

The estimate-at-completion figures reflect the budget-to-date amount and overrun or underrun for each task. The estimate-at-completion indicates what you will have spent on the task when 100 percent complete, having performed at the current pace.

Take Perform B in **Exhibit 6.6-4**. This task has an overrun of $1,038. To determine the estimate-at-completion, you add the $1,038 to the $3,950 to acquire an expected cost of $4,988 once the task is 100 percent complete. Hence, you add the budget to the overrun to obtain the estimate-at-completion.

Under some circumstances, you will have an underrun as shown for activity Perform F in **Exhibit 6.6-4**. To calculate the estimate-at-completion, subtract the underrun from the budget amount (subtract $482 from $5,804 to obtain $5,322).

After calculating the estimate-at-completion costs for each task, you can sum them to derive the figure for the entire project.   This amount is called **management-estimate-at-completion (MEAC)**, and it reflects what you will have spent once the project is 100 percent complete, assuming you continue to perform at the current pace. The formula for the MEAC is:

MEAC = $\sum$ estimate-at-completion for each task

The difference between original budgets to complete the project and the MEAC indicates a variance. If the MEAC exceeds the budget amount, you may have a problem unless you have access to unlimited funds. The goal is to institute measures to reduce costs and consequently the MEAC; otherwise, your project costs will exceed expectations and reduce your profit margin.

Management Reserve

Some project managers employ a management reserve to help them adjust to circumstances when actual costs exceed budget. The management reserve is a percentage of the estimate to complete the project (typically 5 to 7 percent). Thus, take the estimate of $48,219 to complete the project shown in Exhibit 6.6-4. The project manager might allow a portion to serve as a contingency fund to use in emergencies, such as special training or overrunning a task. The project manager might take, for example, 7 percent of the $48,219 ($3,375) and use that amount to address an emergency. The revenue requirements for the project would then be $51,594 rather than $48,219 ($48,219 + $3,375).

Project managers have discretionary power over the management reserve. They can use it for whatever they deem necessary (within legal constraints). It serves as an effective tool for recovering from unexpected costs.

Conclusion

Money is the energy that sparks a project. Nothing occurs until someone authorizes expenditures. Money is a finite resource, so you must carefully determine its disposition. You also need to track its usage to ensure that waste is limited, or better yet, nonexistent.

30600 COST CALCULATION

P² M² CYCLE

Resources	Management Direction	Cost Information	Work Breakdown Structure	Time Estimates	Statement of Work	Risk Identification	Revised Schedule	Major Participants	TASKS	Project Manager	Senior Management	Project Sponsor	Project Team	Client	Procedures	Management Reserve	Cost Estimates
	X							X	**30605** Assign responsibilities for budgeting	X					X		
								X	**30610** Determine methods to classify costs	X					X		X
	X	X				X	X		**30615** Determine when to perform budgeting	X					X		
	X							X	**30620** Establish a procedure for budgeting	X					X		
	X					X			**30625** Ensure budget complies with managerial direction	X	X				X	X	X
X		X		X					**30630** Determine the resources that will consume the most expenditures	X	X		X				X
X		X	X	X		X	X		**30635** Estimate project cost per task	X							X
X		X	X	X			X		**30640** Estimate total cost per resource	X							X
X				X					**30645** Determine direct and indirect costs	X		X	X				X
	X			X		X	X		**30650** Determine the amount to set aside for management reserve	X	X					X	
X	X	X	X	X	X	X	X	X	**30655** Determine the cumulative estimates for labor, training, equipment, supplies, facilities, information, materials, travel and other items important for completing the project	X							X
X	X	X	X	X	X	X	X	X	**30660** Estimate total cost for the entire project	X						X	X
									MEASURES OF SUCCESS								
									Does the budget provide reasonable confidence in achieving project goals and objectives?								X
									Does the budget reflect constraints identified in the statement of work?								X
									Have all the major participants given their concurrence to the budget?							X	X
									Is the budget based upon the WBS, time estimates, schedule and resource allocations?							X	X
									Are the budgeting requirements for each task determined?								X
									Have the monies been authorized?							X	X
									Can the budget be calculated differently?								X
									Is the budgeting process documented?						X		

7

Project Organization (40000)

Introduction

While some people construe the word "documentation" to mean red tape or bureaucracy, they usually admit that it is an important factor in the success of any project. Without adequate documentation, the productivity of a project declines.

Too much documentation can jeopardize a project as much as too little documentation. The idea is to have the right documentation. This may include memos, project procedures, work flows, project manual, project history files, and project newsletters.

You will need one common location to store most of this documentation—the project library.

40000 PROJECT ORGANIZATION

Newsletter	System Manuals	Miscellaneous Documentation	Selected Project Management Package	Administrative Requirements	Project Objectives	Project Goals	Access Control	P² M² CYCLE / TASKS – MEASURES OF SUCCESS	Project Manager	Senior Management	Project Sponsor	Project Team	Client	Project Office	Newsletters	Memos	Project Library	Reports	Forms	Workflows	Project History Files	Location	Procedures	Team Structure	Responsibility Matrix	Organization Chart	Project Manual	Support	Automated Project Management
								TASKS																					
								40100 Automated project management	■		■		■															■	■
								40200 Team organization	■		■	■	■											■	■	■			
								40300 Project procedures	■		■	■	■										■						
							■	40400 Project history files	■		■										■	■	■						
								40500 Workflows	■		■	■								■									
					■	■	■	40600 Forms	■		■								■										
			■		■	■	■	40700 Reports	■	■	■	■	■					■											
■	■	■			■			40800 Project library	■		■						■												
								40900 Memos	■	■	■	■	■			■												
				■				41000 Newsletters	■		■	■	■		■														
			■		■			41100 Project office	■		■			■															
■					■			41200 Project manual	■		■	■															■		
								MEASURES OF SUCCESS																				
								Is project management software selected?																				■	■
								Is there an effective team organization?																■	■	■	■		
								Are project procedures available?						■		■							■				■		
								Are project history files available?								■					■	■							
								Are workflows available?						■		■				■			■				■		
								Are useful forms available?											■								■		
								Are the right reports being produced?						■		■	■	■									■		
								Is a project library available?								■						■							
								Are memos prepared when appropriate?							■												■		
								Is a newsletter published?						■							■						■		
								Does a project office exist?						■															
								Are project manuals published?																			■		

INPUTS

	Cost Estimates	Contingency Plans	Changed Schedule	Changed Budget	Bar Chart	Organizational Standards	Workflows	WBS	Management	Allocated Resources	Vendor Profile	Constraints	Tool Information	Statement of Work	Major Participants	Storage Requirements	Facilities	Updated Schedule	Updated Reports	Updated Budget	Time Estimates	Risk Identification	Revised Schedule	Resource Histograms	Project	Procedures	New Schedule	New Budget	Network Diagram	Impact Analysis	Responsibilities	Matrices	Reports	Organization Chart	Memos	Forms	Bar Chart	Project Library	Project History Files	Project Activities & Events	Project Manual
											■	■	■	■	■																											
							■	■	■					■	■																											
						■		■	■						■																											
	■	■	■	■	■		■	■						■	■	■	■	■	■	■	■	■	■	■	■	■	■	■	■	■	■										
						■			■					■	■									■																	
									■						■																										
								■						■	■																										
									■						■																										■
						■			■					■																											
								■						■																									■		
																																■	■								■
	■							■						■					■				■		■	■	■	■	■			■		■	■	■	■	■				

7.1 Automated Project Management (40100)

For many years, project managers used only large computing systems—usually minicomputers and mainframes—to operate **project management (PM) software**. Today project managers have project management software that runs on personal computers (PCs). These products are affordable and powerful tools that every project manager should consider using.

Getting Started with PM Software

If you use project management software for your project, you have to make some important decisions concerning your workstation setup. Depending on your requirements, you will need most or all of these items: personal computer; plotter and/or printer; software; manuals; and the appropriate supplies and furniture.

The cost and capability of these items can very dramatically. The key is to make an appropriate technology investment based on the requirements of your project.

Some Key Insights

If you elect to use project management (PM) software, remember these very important concepts. Using the software will involve a learning curve. No PM software is so user-friendly that a person can sit at a personal computer and start using the package. Team members must learn how to activate certain functions and the meaning of different fields, icons and processes.

Learning this information takes time. The more your team will rely on the package, the longer the learning curve will last. If you do not provide enough time for training, you may experience hostility by team members towards the package, including the desire to avoid using the software.

Prior to purchasing a package, you must define your requirements by determining the features needed in your PM software. Some project managers purchase software and then discover that the package does not meet their needs.

It is critical for project managers to understand that PM software will not manage the project for you. Only you can manage the project. The software will not make decisions for you, let alone the right decisions. The software is nothing more than a tool. How you employ it will substantially impact the success of your project.

Many artists use a paintbrush and a palette. Yet, not all artists are good painters. Successful painters know **how** to use the paintbrush and mix colors. In a similar vein, project managers will not be successful simply by using a PM package. They must know how to use it under the right circumstances.

Project management software does not replace the need to build a good work breakdown structure (WBS), define schedule logic, make realistic estimates, and develop responsibility matrices. You must still perform the necessary work leading to a useful project plan. No software package will do these tasks for you. What the software does for you is take the data you provide and process it into useful information.

A final insight is that your package should have obsolescence protection. Buying a package and not receiving upgrades can result in a long-term decline in productivity. As your needs expand, you may want, even need, to incorporate any upgrades to the package. Without this capability, you may find it necessary to purchase a new system when an upgrade of the existing system would have sufficed.

How to Select Project Management Software

You should conduct a thorough requirements analysis prior to buying your software. The following procedure can help you make the appropriate decision.

Requirements	Assigned Value
Create a project	2
Duplicate a project	1
Modify a project	0
Handle projects > 50 activities	2
Choose between ADM and PDM	1
Provide standard reports	2
Develop custom reports	1
Enter logic relationships among activities	1
Generate resource histograms	2
Perform costing	2
Conduct "what-if" analysis	0
Check for network logic errors	0
Perform calendaring	1
Choose between batch and interactive modes	1
Provide import and export capabilities	0

Legend: 2 = necessary requirement, 1 = wanted but not needed,
0 = not wanted or needed.

DETERMINATION OF PROJECT MANAGEMENT SOFTWARE REQUIREMENTS
EXHIBIT 7.1-1

1. List your requirements and determine what project management capabilities are required. Then find out what the PC hardware standard is for
   your organization.
2. Assign a weight to each requirement to signify its relative level of importance. See **Exhibit 7.1-1** for an illustration of this step.

Requirements	Assigned Value	Package #1	Package #2	Package #3	Package #4
Create a project	2	2	2	2	2
Duplicate a project	1	1	0	0	0
Modify a project	0	0	0	1	0
Handle projects >50 activities	2	1	2	2	1
Choose between ADM and PDM	1	0	1	0	0
Provide standard reports	2	0	0	0	0
Develop custom reports	1	1	0	1	1
Enter logic relationships between activities	1	0	1	0	1
Generate resource histo-grams	2	2	0	2	0
Perform costing	2	2	2	0	1
Conduct "what-if" analysis	0	0	0	1	0
Check for network logic errors	0	1	1	0	0
Perform calendaring	1	0	0		0
Choose between batch and interactive modes	1	1	1	1	0
Provide import and ex-port capabilities	0	0	0	0	1
Total Score	16	11	10	10	7

Legend: 2 = meets necessary requirement, 1 = meets wanted but not needed requirement, 0
= does not meet wanted or necessary requirement.

SELECTION OF THE
RIGHT PROJECT MANAGEMENT PACKAGE
EXHIBIT 7.1-2

3. Analyze and score each product in terms of the requirements you have listed (see **Exhibit 7.1-2**). You can determine whether the software meets certain requirements by interviewing salespeople for the product, reading sales literature (with skepticism), reviewing technical comparison reports, and talking with people who have used the software.
4. Select the top-scoring packages for further investigation.
5. Conduct a demo of each package either by yourself or with your team members. During these demos, pay particular attention to the "value-added" features of the product. Some typical value-added benefits include user-friendliness and responsiveness to requests. Although such benefits are difficult to quantify, give them the utmost attention.
6. Select the package that best meets your needs.
7. Train people to use the software package. Recognize that people will have a learning curve to overcome. You will need to provide the necessary time and money to enable them to acquire a desired level of expertise in using the package.

Necessary Features of a PM Package

A PM software package must have certain features. Building a project is critical. So is maintaining and duplicating a project. Duplicating a project enables using the previous schedule logic and estimates for a similar project. This capability helps to avoid re-entering information into the computer. You can then modify the duplicate copy to meet the requirements of the present project.

Allowing a large number of activities per project is important, too. Most packages vary in this capability. Some allow only 250 activities per project, while others allow up to 10,000. Packages claiming unlimited activities per project allow you to create several subprojects and connect them. Before purchasing a package, determine the maximum number of activities you will require for your project.

Being able to select the arrow diagramming method (ADM) or the precedence diagramming method (PDM) is an important feature. You may want to use arrow diagrams for your schedules. Or, you may desire a precedence diagram. Regardless, many packages give you a choice. Project managers use ADM for construction projects and PDM for projects in service industries, such as information systems.

Producing specialized and standardized reports is essential. You should have the capability to select and print information based on specific criteria. Also, you should be able to mask, or hide, information on standard reports. This masking feature prevents people from seeing certain information. It also prevents overloading team members, senior management, and the client with too much information.

Developing a **target schedule** and a **current schedule** is another important feature. The target, or baseline, schedule is the one you agree to follow. The current schedule is a combination of the target schedule and status against the items contained therein. The current schedule also projects when current and future activities will start or finish, or both, if the project continues at its present pace. Project managers pursue whatever is necessary to match the current schedule with the target schedule. Incidentally, most packages allow you to modify the target schedule whenever necessary.

Your package should allow entering three basic logic relationships between activities: finish-to-start; start-to-start; and finish-to-finish. Some packages permit only one logic relationship between activities, usually finish-to-start; however, most allow all three types of relationships.

A package should also enable assigning one or more resources to an activity. These resources can be either labor or nonlabor. For labor, most packages will permit assigning a specified number of hours per day that a person can work on a task. For nonlabor resources, you can assign equipment and supplies to a task.

If you can assign resources, then your package should enable generating resource histograms. Most packages will, according to your specifications, provide you with options to generate histograms for one person, for a select group of people, or for the entire team.

In addition to building histograms, your software package should enable resource leveling. Most packages vary on the criterion for leveling a histogram. Some packages level by changing the start and finish dates of activities. Other packages automatically select the relationship type (e.g., finish-to-start) between two or more activities in a way that best levels the histogram. Most packages, regardless of whether changing the duration for activities or selecting the appropriate relationship type between two activities, will extend the project completion date. Only a few software packages attempt to level a histogram without extending the project completion date.

Your package should permit developing other graphics, not just histograms or network diagrams. With a graphics software package, you can generate meaningful charts based on the data stored in the system's memory. Some typical graphics include pie charts and cumulative curves (also called "S" curves).

A package should also enable costing. Most packages have this capability, although they vary in capability. Typically, you can develop three sets of costing data: cost estimates; budget costs; and **estimate-at-completion (EAC)**. Estimates are cost data you expect to spend on each activity and for the entire project. Budget costs are "frozen" estimate costs. The concept is the same as a target schedule. You create a baseline using your cost estimates, thereby committing yourself to completing each task for a certain cost and doing the same for the entire project. The concept of EAC is similar to the current schedule. The EAC indicates how much you have spent for the entire project up to a specific point in time. It also tells how much each task and the entire project will cost to complete if the project continues at its current pace (i.e., cost projections).

In addition to costing, your PM package should allow you to perform "what-if" scenarios. You should be able to change specific variables, such as the duration of a task, and have the software show you the impact those alterations have on project costs and schedules. Such capabilities permit you to experiment and assess the impact of a change. For instance, if you extend the duration of a task you can answer questions such as:

> How will it impact costs?
> How will it affect subsequent tasks on the paths?
> What is the effect on the critical path?
> Will the project completion date need renegotiation because of the change?

The capability to perform what-if scenarios enables you to determine the effect of a change before the project begins, instead of after it is implemented.

Your package should also check for network logic errors. Occasionally, project managers inadvertently connect tasks together illogically. For instance, a task designated as a successor is mistakenly made a predecessor to an activity that was supposed to start earlier. When the software does the forward and backward computing passes to calculate early and late dates, it will discover the error and cease calculating. You can then correct the mistake, and the software will repeat the calculations.

Your package should also allow you to develop summary reports regarding schedules, costs, and resource utilization. These reports give subtotals and grand totals. However, many more sophisticated packages have the capability to develop summary reports based upon different levels within the WBS. Summarizing at different levels in the WBS gives you the opportunity to develop meaningful and useful reports to senior management and your client. A term often used to describe summary schedules in the form of bar charts is **roll-up schedule**.

You should also have the capability to do calendaring. The software should permit changing the work week for the entire project. For instance, you and your project team may have a work week of 4 or 6 days. Most packages assume a 5-day work week but enable you to change this. You can also designate nonwork days besides the standard weekends and statutory holidays. Some packages also enable designating work weeks and nonwork weeks for specific labor resources.

Your package should allow you to access project data and produce reports. Periodically, you will enter status data in regards to the schedule. After entering the data and generating the necessary information, you may want to save versions of the project data for future reference. If your package does not have this capability, you can overcome this shortcoming by duplicating the files and saving them under a different file name.

Your package should give you the option of entering data in either the batch or interactive mode. Batch means building a large data file and importing it into the package for later use. Interactive means entering each datum one by one into the system. For large projects, the batch mode is far more preferable than the interactive mode because you can quickly build a large file and "dump," or import, it into the PM software. In the interactive mode, you enter each datum field-by-field and screen-by-screen, which can prove to be long and tedious. You should be able to select between batch and interactive modes for both building the schedule and entering status information.

Finally, your PM package should have the capability to interact with other software packages. For instance, you should be able to import data from and export data to a database package. In addition, you should have the capability to import data from and export data to another PM package. Importing or exporting data between packages usually requires creating a "flat" or "meta" file.

Further Considerations When Purchasing a PM Package

When performing your requirements analysis, you must do more than just research the capabilities of the software. You must look at vendor service, too. Prior to purchasing the software, compile questions for the vendor. This list should include such questions as:

- Are upgrades free, or is there a fee?
- Does the vendor distribute updates to customers?
- Does the vendor offer users free service for answering direct queries?
- Does the vendor provide on-site assistance of any kind? If so, is the assistance free?
- Does the vendor publish a newsletter about the package?
- Does the vendor sponsor user groups?
- Does the vendor provide satisfactory documentation on how to use the software?
- Is the training free, or is there a fee?
- What is the overall reputation of the vendor?
- Will the vendor provide training?

Other Project Software

Project managers use other PC software related to project management, including packages for word processing, spreadsheets, graphics, database management, and communications.

- **Word processing.** Allows creation of narrative text, such as project detail and summary reports and procedures and enables you to draft, revise, and archive documents you develop.
- **Spreadsheets.** Enable displaying information in tabular form. Spreadsheets show information via rows, columns, and cells and enable you to develop algorithms to calculate data.
- **Graphics.** Allow creation of diagrams and charts to display data. Some common graphics are bar charts, pie charts, histograms, trend charts, cumulative curves. graphics packages communicate information clearly and concisely.
- **Database Management.** Enables you to store data and arrange it according to a logical schema you develop. The data then serves as a basis for developing reports and charts.
- **Communications.** Communications packages enable transmission of data via a modem or over a local area network from one workstation to another. Communications software allows you to acquire data from places located over a wide geographical area and within a short time period.

Conclusion

The worst approach is to purchase a PM software package on impulse. The best approach is to first determine your exact requirements. The wrong decision can mean not only a poor investment in hardware and software but also a decline in productivity.

40100 AUTOMATED PROJECT MANAGEMENT

INPUTS					P² M² CYCLE	RESPONSIBILITIES						OUTPUTS
Vendor Profile	Constraints	Tool Information	Statement of Work	Major Participants		Project Manager	Senior Management	Project Sponsor	Project Team	Client	Support	Selected Project Management Package
					TASKS							
			■	■	**40105** Determine the requirements of the software	■		■		■		■
		■			**40110** Determine the platform it will run on	■						■
	■				**40115** Determine the amount of money you are willing to spend	■						■
		■			**40120** Conduct inventory of current hardware	■		■				■
		■			**40125** Conduct inventory of current software	■		■				■
	■	■			**40130** Determine what additional hardware is needed	■		■				■
	■	■			**40135** Determine what additional software is needed	■		■				■
■					**40140** Ensure vendor support is adequate	■					■	■
■					**40145** Ensure user documentation is available	■					■	■
■		■		■	**40150** Conduct a demo of the software before purchase	■						■
■		■		■	**40155** Conduct a review of tools	■			■			■
■	■	■		■	**40160** Select the tool	■			■			■
				■	**40165** Train everyone on the software	■			■		■	■
					MEASURES OF SUCCESS							
					Have all the software packages for the project been determined?							
					Have all the major requirements for each package been identified?							■
					Is training available to use the package?						■	■
					Has the quantity of packages been determined?							■
					Has a list of the people needing and using specific software been created?							■
					Have all the major users of the software provided feedback on the selected software packages to review?							■
					Is the software accompanied with good user documentation?						■	■

7.2 TEAM ORGANIZATION (40200)

The purpose of establishing a team organization is to maximize the use of human resources on a project. In theory, through organization comes efficiency and effectiveness.

Two basic project team organizations exist. These are the **task force** and **matrix structures**.

The task force occurs where a group of individuals work full-time on a project (**Exhibit 7.2-1**). Their attention is devoted entirely towards achieving the overall project goal. This structure often works best for nonrepetitive types of projects that occur for short periods of time. It is also best for projects where the required skills are not scarce.

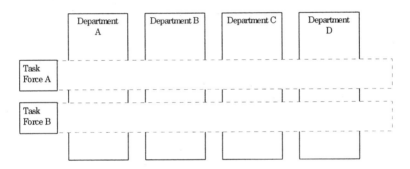

TYPICAL TASK FORCE STRUCTURE
EXHIBIT 7.2-1

The matrix structure occurs where individuals support multiple projects while working for a functional manager (**Exhibit 7.2-2**). This structure often works best for projects requiring the use of scarce technical skills and limited staffing. These project can last for any duration.

To develop an efficient and effective team organization, consider the following:

1. Maintain span of control—try to keep group size to 10 or fewer.
2. If you must exceed 10, then break the group into subgroups and appoint leads. This will allow you easier control over the team.
3. Construct an organization chart showing both the vertical and horizontal reporting relationships. Publish the chart so that everyone knows the reporting structure and their respective areas of responsibilities. **Exhibit 7.2-3** contains a typical project organization chart.

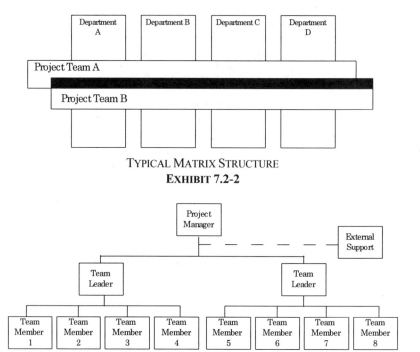

TYPICAL MATRIX STRUCTURE
EXHIBIT 7.2-2

TYPICAL PROJECT ORGANIZATION CHART
EXHIBIT 7.2-3

Publish a **responsibility matrix (Exhibit 7.2-4)**. This chart shows the tasks to be performed on the horizontal axis and the people to perform those tasks on the vertical. The matrix includes notation for recording the resources performing (P), reviewing (R), approving (A), contributing (C) and informing (I).

	Perform A	**Perform B**	**Perform C**
Sue Abrahms	A	A	A
Mike Epp	P	I	R
Gary Jones	I	P	
Karen Ryder	C	R	
Noel Smith			C
Yvonne Su	R	C	P
Wendy Wong	R		I

TYPICAL RESPONSIBILITY MATRIX
EXHIBIT 7.2-4

40200 TEAM ORGANIZATION

	INPUTS				P2 M2 CYCLE	RESPONSIBILITIES					OUTPUTS		
Work Breakdown Struct	Statement of Work	Major Participants	Management Direction	Allotted Resources		Project Manager	Senior Management	Project Sponsor	Project Team	Client	Team Structure	Responsibility Matrix	Organization Chart
					TASKS								
	■	■	■	■	**40205 Determine the type of organizational structure**	■		■		■	■		■
		■	■	■	**40210 Develop and publish an organization chart**	■		■	■	■			■
			■	■	**40215 Ensure effective span of control**	■					■		
■					**40220 Prepare and publish a responsibility matrix**	■			■	■		■	
					MEASURES OF SUCCESS								
					Is the most appropriate organizational structure being used?						■		
					Is an effective span of control in place?						■		
					Has an organization chart been developed?								■
					Has the organizational chart been distributed?								■
					Has everyone performing tasks received a copy of the responsibility matrix?							■	

7.3 PROJECT PROCEDURES (40300)

Having formal **project procedures** in place on a project is important. Unfortunately, many procedures go undocumented. The result is that no one follows them because they have no knowledge about the procedures. As a project manager, you not only want to but **need** to develop and enforce good procedures that cover several topics. At least three good reasons exist for having written procedures.

1. They improve communications. People can refer to the documents to obtain vital information to improve their performance.
2. They bring everyone onto the same wavelength. Well-written procedures delineate who does what, and even why. Instead of everyone going in different directions, they can all move in the same direction.
3. They improve productivity. By having access to well-written procedures, team members can refer to the appropriate procedure for information rather than waste time searching for it or interrupting other people.

Naturally you will want to avoid developing procedures for every single action on the project. Procedures for every action would be counterproductive and lead to overcontrol and stifling the team's creativity. The procedures should cover only the essential topics, not every subject. The project manager determines what topics are essential and the appropriate level of detail.

Procedures should be developed for:

*   Change control
*   Equipment utilization
*   Use of forms, including who completes and approves them
*   Organizational structure (such as an organizational chart reflecting reporting relationships)
*   Responsibilities of project participants
*   Schedules
*   Special meetings (i.e., status review and quality reviews)
*   Material purchases
*   Supply purchases

Each procedure should answer questions regarding the who, what, when, where, how, and why for each specific subject. In addition, each procedure should have these sections:

*   Purpose
*   Scope
*   Contents
*   Approval (the signature of the project manager)
*   Appendices (any supplemental material)

The purpose is a short summary explaining what the procedure will cover and why. It should not exceed more than three or four sentences. The scope describes the parameters of the procedure. The contents section contains descriptive information. It tells "what" and "how" and communicates the information in one of three formats: narrative; sequential; or item-by-item.

The **narrative format** describes the "what" and "how to" of a procedure, as in an essay. You should use complete sentences, as shown in **Exhibit 7.3-1**.

Completing the Activity Estimating Form (AEF) #1

This procedure describes how to complete the Activity Estimating Form (AEF) #101. For processing the form, refer to procedure AEF #102, Processing the Activity Estimating Form. For each work package level item in your work breakdown structure (WBS), estimate the time required to complete it. Be sure to complete the following for each field: (1) activity number; (2) activity description; (3) most optimistic time; (4) most likely time; (5) most pessimistic time; (6) expected time; (7) productivity adjustment percent; (8) revised expected time; (9) clock hour divisor; and (10) duration. What now follows is a short description of each field.

Field Name	Description
Activity Number	Numerical designation uniquely identifying the task
Activity Description	Short narrative description of the task
Most Optimistic	Number of hours to complete the task under ideal conditions
Most Likely	Normal/average hours to complete the task
Most Pessimistic	Number of hours to complete the task under the worst conditions
Expected Time	Expected hours to complete the task using the PERT formula
Productivity Adjustment Percent	Percentage of nonproductive time employee(s) expend when working on the task
Revised Expected Time	Percentage in column 7 multiplied by the figure in column 6
Clock Hour Divisor	Number of hours per day the employee(s) will work on the task
Duration	Number of days the task will take to finish; derived by dividing the figure in column 9 into the one in column 8

COMPLETING THE ACTIVITY ESTIMATING FORM (AEF) #101
NARRATIVE FORMAT
EXHIBIT 7.3-1

The **sequential format** describes the "what" and "how" to information outline format. You remove the subjects in each sentence and use indentation to show the operations, tasks, subtasks, and steps to implement the procedure. An example of a procedure using this format is shown in **Exhibit 7.3-2**

Completing the Activity Estimating Form (AEF) #101

This procedure describes how to complete the Activity Estimating Form (AEF) #101.

  A. Obtain a copy of AEF #101 from your areas form file.
  B. For each work breakdown structure (WBS) element, enter the applicable number below:

   1. Numerical designation uniquely identifying the task
   2. Short narrative description of the task
   3. Number of hours to complete the task under ideal conditions
   4. Normal/average hours to complete the task
   5. Number of hours to complete the task under the worst conditions
   6. Expected hours to complete the task using the PERT formula
   7. Percentage of nonproductive time employee(s) expend when working on the task
   8. Percentage in column 7 multiplied by the figure in column 6
   9. Number of hours per day the employee(s) will work on the task
 10. Number of days the task will take to finish; derived by dividing the figure in column 9 into the one in column 8

COMPLETING THE ACTIVITY ESTIMATING FORM (AEF) #101
SEQUENTIAL FORMAT
EXHIBIT 7.3-2

The **item-by-item format** describes a series of topics related only by general subject matter. It may contain blocks of information that differ from one another. **Exhibit 7.3-3** is an example of a procedure using the item-by-item format.

You must judge which format to select. Some project managers prefer narrative descriptions, while others prefer easier-to-read formats, such as sequential and playscript. Whichever format you choose, ensure that the resulting procedures are presented with plenty of white space to improve readability, emphasize the use of graphics, and write clearly and concisely.

Completing Project Forms

This procedure describes under what circumstances you want to complete these forms: (1) Activity Estimating Form; (2) Record of Estimates to Complete Work Package Level Items; and (3) Record of Data for Building a Resource Histogram for Labor.

I. Activity Estimating Form (AEF #101)

    1. Complete when making estimates for each work package level item in the WBS.
    2. Refer to AEF procedure #101 for instructions on how to complete this form.
    3. Refer to AEF procedure #102 for disposition of the completed form.

COMPLETING PROJECT FORMS
ITEM-BY-ITEM FORMAT
EXHIBIT 7.3-3

Besides considering format and appearance of procedures, you must wrestle with some very important administrative considerations such as:

- How many copies should be produced?
- How often should the procedures be reviewed?
- Should everyone have a copy?
- What topics should the procedures cover, and at what level of detail?
- Who, in addition to yourself, must approve them?
- Who should prepare the procedures?
- Who will maintain them?

Conclusion

Procedures are essential for successful project management. Yet, maintaining good procedures is like performing a high-wire act. Highly detailed procedures eliminate freedom of action on the project. Ill-defined, vague procedures do nothing to improve the circumstances. You want procedures that project participants will find useful but not inhibiting.

40300 PROJECT PROCEDURES

INPUTS				P² M² CYCLE / TASKS	RESPONSIBILITIES					OUTPUTS
Organizational Standards	Workflows	Major Participants	Management Direction		Project Manager	Senior Management	Project Sponsor	Project Team	Client	Procedures
■	■			**40305** Determine type of procedures	■					■
		■		**40310** Determine who should prepare the procedures	■		■			■
			■	**40315** Determine topics	■					■
■				**40320** Determine format	■					■
■		■	■	**40325** Determine who must approve them	■	■			■	■
■		■	■	**40330** Determine who will maintain them	■		■			■
■				**40335** Determine how often to review	■					■
		■		**40340** Determine number of copies to produce	■			■	■	■
		■	■	**40345** Determine who should get copies	■			■	■	■
	■			**40350** Determine level of detail	■					■
■			■	**40355** Determine storage medium	■					■
		■		**40360** Distribute procedures	■			■	■	■
				MEASURES OF SUCCESS						
				Are the topics for the procedures defined?						■
				Is the format of the procedures defined?						■
				Are the responsibilities for creating, reviewing, approving, distributing and maintaining procedures determined and communicated?						■
				Are the procedures at the appropriate level of detail?						■

7.4 PROJECT HISTORY FILES (40400)

Project history files provide team members with a wealth of useful information. These are files containing information about your project, from budgeting to responsibilities.

Specifically, one or more files are maintained for each major topic, including:

- Correspondence with management and client
- Drafts of all significant documentation
- Estimates
- Forms
- Memos
- Minutes of meetings
- Procedures
- Reports
- Responsibilities
- Schedule updates
- Types and versions of schedules
- Work breakdown structures

Well-maintained project history files offer several advantages. First, they enable good traceability. Whenever you must ascertain the cause of a problem, you can review the files for information. This will help you to identify the cause and fix the problem.

Project history files also facilitate audits. Occasionally, senior management may conduct an audit and request specific information. The history files serve as an excellent reservoir of information to respond to such inquiries.

Project history files also protect the project manager. If something goes awry, for example, as a result of a client's decision, you can produce any relevant paperwork from the files to substantiate your defense.

Reports can be more easily developed from history files. You can directly access the files and extract the necessary information to build the reports, thereby eliminating endless searches. –

Project history files also enable shortening the learning curve of new team members. You can have them review the files, thereby providing the opportunity for new team members to become productive more quickly.

The files are also a convenient mechanism for storing project documentation. Rather than having documentation hidden in people's desks, the paperwork is stored in a central location accessible to everyone. People will not waste time locating documentation.

Finally, history files improve communications. People can refer to the files to clarify any ambiguities or to find answers to questions.

Administrative Issues

To reap the benefits of having project history files, one needs to address two important administrative issues. First, determine who will establish and maintain the files. You must appoint someone as the custodian of the files. This person's responsibilities should include setting up the files, adding or removing documentation, and regulating who has access.

Building the files is easy. The custodian simply uses a large filing cabinet and fills it with folders. Each folder has a label describing its contents. Second, select a central location for the files. Be sure to position the files so team members have easy access.

Another approach is electronic file sharing. A common file server with a predesignated directory (structure) can contain all the project's history files. Team members would be granted read-only access. Another option is a home page on your firm's intranet, with links to documentation that is referenced.

Many project managers do not take the effort to establish project history files. They claim that it takes too much time and effort initially to establish and maintain project history files. Ironically, time and effort expended early save more time and effort later in the project.

40400 PROJECT HISTORY FILES

P2 M2 CYCLE — INPUTS · TASKS · RESPONSIBILITIES · OUTPUTS · MEASURES OF SUCCESS

TASKS
- 40405 Determine contents
- 40410 Determine who will establish and maintain the file
- 40415 Select central location of files
- 40420 Determine level of access to files
- 40425 Determine backup of files

MEASURES OF SUCCESS
- Are the contents of the files defined?
- Does a file checkout procedure exist?
- Are the files located in an accessible location?
- Are the contents of the files complete?
- Are the contents of the files current?
- Are the responsibilities for creating and maintaining the files defined and communicated?
- Is the setup and use of project history files documented?

	40405	40410	40415	40420	40425	Contents defined?	Checkout procedure exist?	Accessible location?	Contents complete?	Contents current?	Responsibilities defined/communicated?	Setup/use documented?
RESPONSIBILITIES												
Project Manager	▨	▨	▨	▨	▨							
Senior Management												
Project Sponsor												
Project Team		▨										
Client												
OUTPUTS												
Procedures	▨	▨	▨	▨	▨	▨	▨				▨	▨
Location			▨					▨				
Project History Files	▨	▨	▨	▨	▨	▨			▨	▨		
INPUTS												
Bar Chart	▨											
Changed Budget	▨											
Changed Schedule	▨											
Contingency Plans	▨											
Cost Estimates	▨											
Impact Analysis	▨											
Network Diagram	▨											
New Budget	▨											
New Schedule	▨											
Procedures	▨											
Project Charter	▨											
Resource Histograms	▨											
Revised Schedule	▨											
Risk Identification	▨											
Time Estimates	▨											
Updated Budget	▨											
Updated Reports	▨											
Updated Schedule	▨											
Work Breakdown Struc	▨											
Workflows	▨											
Storage Requirements			▨		▨							
Major Participants				▨								
Access Control		▨										
Statement of Work	▨											

7.5 WORK FLOWS (40500)

Work flows graphically display procedures. Work flows may also supplement written procedures to enhance clarification and understanding.

Exhibit 7.5-1 is an example of a work flow. To use work flows, you need to understand the symbols used.

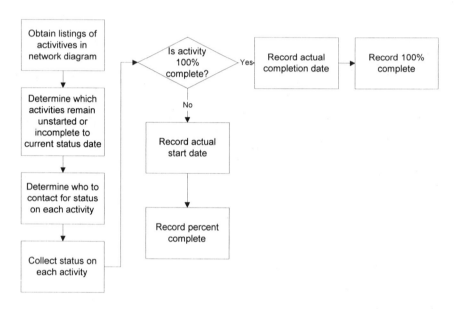

TYPICAL WORK FLOW
EXHIBIT 7.5-1

Process. This is a function that occurs, such as calculate tax. It is represented by a rectangle. Within the rectangle is a verbal description, usually an active verb and an object, of the function.

Decision. This is a point in the work flow where the reader chooses between two or more paths. For instance, you may have to choose whether to calculate your own tax. If you decide "yes," you follow the steps down the appropriate path; "no," then you take another path. You do not always need to show a dichotomous choice.

Delay. This symbol reflects where a time period elapses when no action occurs, such as an approval. The symbol is D-shaped.

Document. This is represented by a rectangle that is curved in the lower left corner. It stands for a document that is being processed, such as being completed or verified for correctness.

Storage. This symbol represents paper or information stored someplace, e.g., in a filing cabinet or computer. An inverted triangle is used to represent storage. An "x" inside the inverted triangle means destruction of the document or the information.

Vector. An arrow represents the direction of flow throughout the work flow. You must follow arrows to understand the logic of the work flow.

The work flow offers two advantages. First, the reader can acquire an overview of the procedure by quickly glancing through the work flow to obtain an appreciation of the sequence of events. Second, work flows are often easier to follow than a narrative procedure.

To realize these benefits, you must address some administrative issues. You must determine who will draft the work flows and who will update them. You must also determine which procedures will be in the format of a work flow rather than a procedure and which procedures will have work flow documentation accompanying them. After addressing these issues, you will find that work flows serve as a useful tool for managing your project.

40500 WORK FLOWS

INPUTS				P² M² CYCLE	RESPONSIBILITIES					OUTPUTS
Organization Standards	Major Participants	Procedures	Management Direction	TASKS	Project Manager	Senior Management	Project Sponsor	Project Team	Client	Work Flows
	▦			**40505 Assign who should prepare the workflows**	▦			▦		▦
		▦	▦	**40510 Determine topics**	▦					▦
▦		▦	▦	**40515 Determine appropriate level of detail**	▦					▦
		▦		**40520 Determine whether to integrate with other documentation**	▦					▦
▦				**40525 Determine the storage media**	▦					▦
▦				**40530 Determine the symbols to use**	▦					▦
	▦			**40535 Determine who approves them**	▦			▦	▦	▦
▦			▦	**40540 Determine how often to review**	▦					▦
▦	▦			**40545 Determine how many copies to produce**	▦			▦	▦	▦
	▦			**40550 Determine who should get copies**	▦			▦	▦	▦
				MEASURES OF SUCCESS						
				Are the topics for the workflows defined?						▦
				Are the symbols for the workflows agreed upon?						▦
				Are the responsibilities for creating, reviewing, approving, distributing and maintaining workflows determined and communicated?						▦
				Are the workflows at the appropriate level of detail?						▦

7.6 FORMS (40600)

On a project, **forms** are a medium used to capture data, store it, and transport it to the appropriate destination. To maximize their value on a project, forms should:

1. *Have a source and destination.* A source is the place where the blank form is stored, such as a drawer in an office or as an electronic file stored in a computer. A destination is the place where the completed form ends up, such as a drawer in an office or as an electronic file stored on a computer.
2. *Contain simple instructions on how to complete it and what to do with it.* The form should be brief and clear.
3. *Be logically organized.* If the form is stored electronically, it should be organized the same way as the image on the screen. If the same code always appears in a field, then it should be printed on the form for consistency.
4. *Contain plenty of white space.* Crowding of information makes reading the form difficult and adds complexity when completing it.
5. *Ask only for relevant information.* Requests for superfluous information waste time and effort, encourage people to avoid completing the form, and add to administrative overhead to maintain that information.
6. *Require minimal effort to complete.* Completion should be quick. If the information requested is relevant, parsimonious, and organized logically, completing the form should occur quickly.

A considerable number of forms can be used on a project. Some typical forms used in either hard copy or electronic form include

- Activity description
- Activity estimating
- Activity status record
- Change control
- Change request
- Employee assignment
- Estimated labor usage per activity
- Estimated resource (nonlabor) costs
- Estimates to complete work package items
- Problem occurrence
- Purchase order
- Record of actual daily labor utilization
- Record of actual resource (non-labor) utilization costs
- Resource histogram

Copies of blank and completed forms can be stored on a personal or network computer, in a drawer, in the project history file, or in a project manual. Data collected on forms can be used to monitor and control progress and enable collecting performance statistics and preparing lessons learned.

Commonly Used Forms

Project managers will utilize a series of forms to manage their projects. The following represent several commonly used forms, with instructions on how to use them.

Actual Resource Utilization Costs Form

Project: [1]
Date: [2]

Date	Resource Description	Resource Type	Activity Number	Actual Unit Cost in $'s	Actual Total Units	Actual Total Cost in $'s
[3]	[4]	[5]	[6]	[7]	[8]	[9]
	Totals:					0

[1] Project name
[2] Date
[3] The calendar date when the entry was made
[4] A narrative description of the resource
[5] The classification of the resource, such as supplies or equipment
[6] The numerical designation of the activity requiring the resource
[7] The actual cost per resource unit
[8] The actual total number of units
[9] The actual total cost of the resource; calculated by multiplying column [7] by column [8]

ACTUAL RESOURCE UTILIZATION COSTS FORM
EXHIBIT 7.6-1

Change Request Form

Project: _____[1]_____

To: _____[2]_____   Date: _____[5]_____

From: _____[3]_____   Request #: _____[6]_____

cc: _____[4]_____

Description of Requested Change: [7]

Potential Impact of Change: [8]

Requested Priority of Change and Why: [9]

_____[10]_____   _____[11]_____
Requester's Signature   Date

_____[12]_____   _____[13]_____   ❑ Approved
Project Manager's Signature   Date

    ❑ Deferred [16]

_____[14]_____   _____[15]_____
Project Sponsor's Signature   Date   ❑ Rejected

Explanation / Comments: [17]

---

[1] Project name.
[2] To whom the change request will be sent.
[3] From the person completing the request.
[4] People who will receive courtesy copies of the change request.
[5] The calendar time when the originator completed the change request.
[6] The number assigned by the project manager that uniquely identifies the change request.
[7] A narrative description of the change—completed by the requester.
[8] A preliminary analysis of the change—completed by the requester.
[9] Preliminary decision by the requester as to the criticality of the request and rationale.
[10] Requester's signature.
[11] Date signed by requester.
[12] Project manager's signature once an action has been agreed to on the change request.
[13] Date of project manager's signature.
[14] Project sponsor's signature once an action has been agreed to on the change request.
[15] Date of project sponsor's signature.
[16] Disposition of change request—check one box.
[17] Remarks by the project manager explaining the rationale behind the decision.

CHANGE REQUEST FORM
EXHIBIT 7.6-2

Employee Assignment Form

Project: __[1]_____
Date: ___[2]_____

Activity Number	Activity Description	Employee Name	Employee Number	Total Hours to Complete	Early Start Date	Early Finish Date
[3]	[4]	[5]	[6]	[7]	[8]	[9]

[1] Project name
[2] Date
[3] The numerical designator uniquely identifying the activity
[4] A short narrative description of the activity
[5] The name of the person(s) working on the activity
[6] The numerical designator uniquely identifying the person
[7] The total hours required to complete the activity
[8] The earliest moment in time to start the activity
[9] The latest moment in time to complete the activity

EMPLOYEE ASSIGNMENT FORM
Exhibit 7.6-3

Problem Occurrence Record Form

Project: [1]_____
Date: ___[2]_____

Date	Problem Description	Priority	Affected Activities	Impact	Person Responsible	Suspense Date	Additional Comments
[3]	[4]	[5]	[6]	[7]	[8]	[9]	[10]

[1] Project name
[2] Date
[3] Calendar date of the entry
[4] A short narrative description of the delay
[5] The assigned priority of the problem, such as high, medium, or low
[6] The activities (recorded by activity numbers) impacted by the problem
[7] A short narrative description of the effect that the problem will have on the project in respect to cost, schedule, and budget
[8] The people assigned to determine the causes of the problem and its resolution
[9] The tentative date set for the problem's resolution to go into effect
[10] Any other remarks that can clarify the nature and source of the problem

PROBLEM OCCURRENCE RECORD FORM
Exhibit 7.6-4

40600 FORMS

INPUTS				P2 M2 CYCLE	RESPONSIBILITIES					OUTPUTS
Administrative Require	Major Participants	Project Objectives	Project Goals		Project Manager	Senior Management	Project Sponsor	Project Team	Client	Forms
				TASKS						
		▦	▦	40605 Determine types of forms	▦					▦
	▦			40610 Assign responsibility for forms development	▦		▦			▦
	▦			40615 Assign responsibility for forms maintenance	▦		▦			▦
▦				40620 Determine reproduction / medium	▦					▦
▦				40625 Determine sources	▦		▦			▦
▦				40630 Determine destinations	▦		▦			▦
		▦	▦	40635 Determine content	▦		▦			▦
▦				40640 Determine outlay	▦		▦			▦
▦				40645 Determine storage requirements of blank forms	▦					▦
▦				40650 Determine storage requirements of completed forms	▦					▦
				MEASURES OF SUCCESS						
				Are circumstances for using forms defined?						▦
				Are the sources and destinations of forms agreed upon?						▦
				Are the responsibilities for creating, reviewing, approving, distributing and						▦
				Do the forms have the appropriate level of detail?						▦
				Are the forms self-explanatory?						▦
				Are the blank copies of the forms easily accessible?						▦
				Are the forms clear and concise?						▦

7.7   REPORTS (40700)

Reports play a critical role in tracking and monitoring progress of—and ultimately controlling—a project. Most often, reports are used to display information about progress in regards to schedule, budget, and quality.

Reports can take many forms. They can be in tabular or graphical form. They can also appear in hard copy or electronic form.

Some examples of reports used on projects include:

- Arrow diagram
- Bar chart
- Histogram
- Precedence diagram
- Predecessor–successor schedule
- Project schedule
- Project state summary
- Resource cost
- Resource utilization to date

Copies of reports should be stored either physically or electronically (or both). Appropriate places to store reports include the project manual and project history files. Storage and retrieval are important for analyzing sources of problems, developing lessons to be learned from, compiling statistics, and providing audit trails.

Commonly Used Reports

The following are a few of the commonly used reports:

Predecessor–Successor Schedule Report

Project: [1]_____
Date: ___[2]_____

Activity Number	Predecessor(s)	Lag	Successor(s)	Lag	Early Start Date	Early Finish Date	Late Start Date	Late Finish Date	Actual Start Date	Actual Finish Date	Float
[3]	[4]	[5]	[6]	[7]	[8]	[9]	[10]	[11]	[12]	[13]	[14]

PREDECESSOR-SUCCESSOR SCHEDULE REPORT
EXHIBIT 7.7-1

Project Cost Report

Project: ___[1]___
Date: _____[2]_____

Activity Number	Activity Description	Percent Complete	Total Budget	Budget -to- Date	Actual -to- Date	Variance	Estimate to Complete
[3]	[4]	[5]	[6]	[7]	[8]	[9]	[10]

[1] Project name
[2] Date
[3] The numeric designation uniquely identifying the activity
[4] A narrative description of the activity
[5] The amount, in percentage, the activity is complete
[6] The amount originally allocated to complete the activity
[7] The amount originally allocated to complete the activity up to a certain time; calculated by multiplying column [5] by column [6]
[8] The amount actually spent to date on the activity
[9] The amount exceeding what was originally budgeted to date; calculated by subtracting column [7] from column [8]
[10] The amount the activity will be completed for at the current pace of work.

PROJECT COST REPORT
EXHIBIT 7.7-2

Problem Report

Project: [1]_____   Date: __[2]_____

Problem Description	Priority	Affected Activities	Impact	Person Responsible	Suspense Date	Resolved? (Y/N)	Date Resolved	Additional Comment
[3]	[4]	[5]	[6]	[7]	[8]	[9]	[10]	[11]

[1] Project name
[2] Date
[3] A short narrative description of the problem
[4] The level of importance of the problem
[5] The activities impacted by the problem
[6] The consequences of the problem on the schedule, budget, and quality
[7] Who is accountable for fixing the problem
[8] The due date for resolving the problem
[9] Yes or No indicating whether the problem has been resolved
[10] Calendar date when the problem was resolved
[11] Any further remarks to clarify the problem

PROBLEM REPORT
EXHIBIT 7.7-3

Project Status Report

Project: _____[1]_____

To: _____[2]_____

From: _____[3]_____

Date: _____[4]_____

RE: Project Status Report: xx/xx/xx to xx/xx/xx

Accomplishments since last report: [5]

Tasks planned for next reporting period: [6]

Budget update: [7]
    Original budget: $_____
    Estimated cost to date:$_____
    Estimated cost at completion: $_____
    Variance: $_____

Schedule update: [8]
    Baseline start: xx/xx/xx   Baseline finish: xx/xx/xx
    Actual start: xx/xx/xx   Estimated finish date: xx/xx/xx

Outstanding issues: [9]

---

[1] Project name
[2] To whom the project status report will be sent (i.e., project sponsor)
[3] From the project manager
[4] Date of status report
[5] List of key accomplishments since the last status report
[6] List of key accomplishments planned for the next reporting period
[7] Budget information, original budget, expenditures to date and expected
    cost to complete
[8] Schedule information, original start and finish dates and actual start date
    along with the expected completion date
[9] Outstanding issues that are impacting (or could impact) the ability of the
    project to complete on time, on budget, to quality standards, etc.

PROJECT STATUS REPORT
EXHIBIT 7.7-4

40700 REPORTS

Forms	Administrative Require	Major Participants	Statement of Work	Selected Project Manag	Project Objectives	Project Goals	Management Direction	TASKS	Project Manager	Senior Management	Project Sponsor	Project Team	Client	Reports
	▦		▦	▦	▦	▦		**40705 Determine types of reports**	▦		▦		▦	▦
		▦						**40710 Assign responsibility for creation / update of reports**	▦		▦			▦
▦					▦	▦		**40715 Determine contents of reports**	▦					▦
	▦						▦	**40720 Determine frequency of reports**	▦	▦	▦		▦	▦
	▦	▦					▦	**40725 Determine outlay of reports**	▦					▦
	▦			▦				**40730 Determine storage requirements**	▦					▦
	▦			▦				**40735 Determine media for reports**	▦					▦
	▦	▦						**40740 Determine distribution of reports**	▦	▦	▦	▦	▦	▦
								MEASURES OF SUCCESS						
								Are the types of reports agreed upon by all major participants?						▦
								Are the contents of the reports agreed upon?						▦
								Are the responsibilities for creating, reviewing, approving and distributing reports determined and communicated?						▦
								Do the reports have the appropriate level of detail?						▦
								Are the reports self-explanatory?						▦
								Are the reports clear and concise?						▦
								Can the reports be generated easily?						▦

7.8 PROJECT LIBRARY (40800)

The **project library** is the place to store your project documentation to protect it and make it accessible to everyone. The project library should contain all relevant project documentation, including

- Company publications (such as policies and procedures)
- Literature (such as trade publications and books)
- Newsletters
- Project history files
- The project manual
- Project policies and procedures
- Technical manuals

The project library serves as a central repository of information. People must know where to find the information rather than search for it. It also serves as an excellent training center to which new employees can go and review the information, thereby reducing their learning curve.

Building a project library does present some obstacles. You must determine who will establish and maintain the library. The best candidates are individuals who are not working on critical path activities.

You also must institute mechanisms to track or trace borrowed material. A convenient method is to have checkout cards for each file or publication. The person removing the documentation completes the card and gives it to the project librarian.

40800 PROJECT LIBRARY

INPUTS						P² M² CYCLE / TASKS	RESPONSIBILITIES					OUTPUTS
Administrative Requirements	System Manuals	Major Participants	Project Manual	Newsletters	Miscellaneous Documentation		Project Manager	Senior Management	Project Sponsor	Project Team	Client	Library Checkout System
						TASKS						
■		■				**40805** Determine who will establish and maintain the library	■					■
■	■		■	■	■	**40810** Determine the contents	■					■
■						**40815** Determine the location	■					■
■						**40820** Institute mechanisms to track or trace who has removed the material	■		■			■
						MEASURES OF SUCCESS						
						Are the contents of the library defined?						■
						Does a checkout procedure exist?						■
						Is the library in an accessible location?						■
						Are the contents of the library complete?						■
						Are the contents of the library current?						■
						Are the responsibilities for establishing and maintaining the library defined and communicated?						■

7.9 MEMOS (40900)

Many project managers fail to realize that they have a powerful tool, the **memo**. This document can prove useful for several reasons.

It provides a record of results. After you have done something significant by yourself or with other people, you can document the achievement in a memo and forward it to the appropriate people. The results are recorded and stored in a secure location.

A memo breeds commitment. For instance, if someone promises something at a meeting, you can document it in a memo and circulate it accordingly. This helps you achieve visibility and makes it difficult for the person to renege on the promise without your consent.

A memo also provides traceability. If you document something in a memo and store it for future reference, you can then find information on a specific subject when you need it. A memo, or a series of memos, can also help you reconstruct the sequence of events during a specific period.

Memos can help address and resolve issues. If an issue requires attention, you can prepare a memo to give it visibility. A memo can even spark discussion on resolving an issue.

Finally, memos are excellent vehicles for communicating with everyone, whether senior management, a client, or the project team. Memos are a cost-effective way to "get the word out."

To realize these benefits, a memo must have certain qualities. Before you read on, compare the poorly written memo shown in **Exhibit 7.9-1** with the well-written memo shown in **Exhibit 7.9-2**.

To: All Concerned   Date: March 10, xxxx

A status meeting will be held on March 26, xxxx at the Corporate Headquarters Building. It will be at 2:30 p.m.

Attached is some material for your review. Give me a call if you have any questions.

EXAMPLE OF A POORLY WRITTEN MEMO
EXHIBIT 7.9-2

To: Robert James   Date: March 10, xxxx
        Stanley Roberts
        Francis Xavier

cc: Shirley Adams
        Dick Barnstad

Subject: Status Review Meeting

On March 26, xxxx, we will hold an ad hoc status review meeting in the Hanford Conference Room located on the 3rd floor of the Corporate Headquarters building. The meeting will be held from 2:30 – 3:30 p.m., and I will chair it.

At the meeting, we will collect status on the major schedules (see attached) for the Inventory Management System. Please review the attachment prior to attending the meeting.

If you should have any questions or comments, please contact me at 123-4567.

Chris Henderson
Project Manager

EXAMPLE OF A WELL-WRITTEN MEMO
EXHIBIT 7.9-1

Clarity is an important quality. The reader should not have to read the text several times or refer to a dictionary to determine the message.

Conciseness makes good sense, too. Wordiness only confuses the reader and breaks concentration. Long sentences and large paragraphs only frustrate the reader.

Directness is another quality of an effective memo. A well-written memo covers the main points immediately. Extensive prose preceding the main points distracts the reader.

Effective memos must be legible. With today's word processing technology, legibility is usually not a problem.

Lastly, a well-written memo has the right structure. It has a date and an addressee section indicating who receives copies. Typically, the addressee section lists two groups of people, the principals and the interested parties who receive courtesy copies (cc). It also contains a subject heading, a narrative section discussing main topics and, a signature block for the author.

By knowing the difference between a well-written memo and an inadequate one, you can develop memos for all suitable occasions. There are many appropriate circumstances for preparing memos.

A good time to prepare a memo is after a meeting, either with one person or a group. The memo is an excellent medium to list accomplishments and any outstanding issues.

Memos are also useful when you want to raise an issue and force its resolution. Sometimes documenting a problem and disseminating it will force people to recognize that a problem exists and that they need to take action to have it resolved. Memos are a powerful mechanism for acquiring visibility on a specific issue or topic.

Clarifying policy is another opportunity for publishing a memo. If a policy is unclear, you can use a memo to request clarification from the appropriate people.

Communicating important information provides another opportunity to distribute a memo. You may want to impart information to selected individuals. A memo is a quick, efficient way to do it.

A memo is useful for scheduling events. After arranging a suitable date, time, and place for a meeting, for instance, you can record information in a memo and send all the applicable people a copy.

Finally, you can use a memo to document an occurrence of an event. Something of importance, for instance, may have occurred that has an impact on your project. You can prepare a memo explaining the circumstances and the consequences. In this case, the memo provides a historical record of the event.

Methods of Distribution

Memos can be distributed in paper or electronic form. Paper memos are particularly effective if the recipients are not on an easily accessible electronic mail system. Memos sent via electronic mail are best in organizations where office automation is implemented and widely used.

Project managers using e-mail should follow the guidelines for memos described above. Too often e-mail users quickly "bang out" their e-mail messages without putting enough thought into them.

40900 MEMOS

INPUTS			P² M² CYCLE	RESPONSIBILITIES					OUTPUTS
Organizational Standards	Major Participants	Management Direction		Project Manager	Senior Management	Project Sponsor	Project Team	Client	Memos

TASKS

Organizational Standards	Major Participants	Management Direction	TASKS	Project Manager	Senior Management	Project Sponsor	Project Team	Client	Memos
▓			**40905** Determine format	▓					▓
	▓	▓	**40910** Determine topics for memos	▓	▓	▓	▓	▓	▓
	▓		**40915** Determine distribution list	▓		▓	▓		▓

MEASURES OF SUCCESS

			MEASURES OF SUCCESS						
			Are the topics for memos determined?						▓
			Is the format of memos defined?						▓
			Are the responsibilities for creating, reviewing, approving, distributing, and maintaining memos determined and communicated?						▓
			Are the circumstances for developing memos determined?						▓
			Is the distribution list for each type of memo determined?						▓

7.10 NEWSLETTERS (41000)

The project **newsletter** is another effective documentation tool. This medium offers several advantages.

Although seldom appreciated, a project newsletter helps to maintain and improve morale. The newsletter is common to everyone and, if well-written, generates excitement over the project.

The newsletter can be an effective communications tool. It not only excites people but informs them. Typical topics include

- Biographies of the client, management, and team members
- Methods and techniques for performing work
- Obstacles (even failures)
- Performance statistics
- Solutions to common problems
- Project successes

Despite these benefits, you will have to confront three major obstacles when instituting a newsletter.

Politics is a major issue. Issuing a newsletter can result in too much visibility, which can prove disastrous, especially if the project is temporarily in trouble. then there is the problem of someone not on the project receiving an issue and becoming upset over the contents; unfortunately this is a frequent occurrence in many large organizations.

The danger of a newsletter becoming a propaganda "rag" for the project manager is another potentially serious problem facing a newsletter. Occasionally, a newsletter becomes nothing more than a written pontification of the project manager.

Finally, you have to address important administrative matters concerning the newsletter, such as

- How fancy should the newsletter be?
- How much time and money are you willing to spend on one?
- How often will it be published?
- What publication method, e.g., desktop publishing, do you want to use?
- What topics will it address?
- Who must review the newsletter prior to publication?
- Who will produce the newsletter?

Publishing a newsletter is not easy. If you are wondering when a newsletter should be published, consider three criteria. Publish a newsletter when the project team's morale is low, if other traditional communications techniques have failed, or when the project is large (more than 20 people).

41000 NEWSLETTERS

INPUTS				P² M² CYCLE / TASKS	RESPONSIBILITIES					OUTPUTS
Administrative Requirements	Project Activities and Events	Major Participants	Management Direction		Project Manager	Senior Management	Project Sponsor	Project Team	Client	Newsletter
TASKS										
■				**41005** Determine who will prepare the newsletter	■		■			■
■				**41010** Estimate how much money and time to spend	■					■
■	■	■	■	**41015** Determine what topics to address	■					■
			■	**41020** Determine how fancy the appearance	■					■
			■	**41025** Determine how often to publish	■					■
			■	**41030** Determine who must review prior to publication	■	■		■		■
■	■			**41035** Determine what publication method to use	■					■
MEASURES OF SUCCESS										
				Are the topics for the newsletter determined?						■
				Is the format of the newsletter defined?						■
				Are the responsibilities for creating, reviewing, approving, distributing, and maintaining the newsletter determined and communicated?						■
				Is the distribution of the newsletter adequate?						■
				Is the newsletter clear, concise, and meaningful?						■

7.11 PROJECT OFFICE (41100)

The **project office** can exist just about anywhere on a project. Generally, the larger the project, the bigger the project office. The project office offers several advantages to the project manager and the project as a whole.

It serves as a central location for providing easy access to the resources of the project. It also serves as a "command and control center" for the project. Most of the major project management activities and decisions occur at the project office. Locating the administrative staff at the project rather than having it scattered all over the project area helps it to perform its functions in an effective and efficient manner.

Finally, the project office is a central location at which to hold key meetings, such as status and checkpoint reviews, and to discuss and resolve major issues. In other words, it is centrally located for everyone to meet.

It is best to institute a project office as soon as possible. Doing so creates an air of legitimacy and importance to the project besides providing current and future project participants with a place to meet. Some of the key items to locate at the project office are

- Automated project management tools
- Project history files
- Project library
- Project manual

One important feature of the project office is the control wall or room. This wall or room contains all the major project documentation for the project and displays it in a visible, accessible way. The wall or room gives visibility to the project, communicates essential information easily, and makes that information accessible. Some items that can appear on a wall or in a control room are

- Bar charts
- Facilities
- Histograms
- Maps
- Network diagram
- Organization charts
- Reports
- Responsibility matrix
- Statement of work
- Statistics

What is placed on the wall or in the room depends on what the project manager and senior management feel are important to give visibility. Creating a control wall or room does not happen in a vacuum. A project manager must consider a number of factors, such as

- Appearance
- Cost (set-up)
- Maintenance
- Outlay
- Size

41100 PROJECT OFFICE

INPUTS					$P^2 M^2$ CYCLE	RESPONSIBILITIES					OUTPUTS
Administrative Requirements	Selected Project Management Package	Project Library	Project History Files	Project Manual		Project Manager	Senior Management	Project Sponsor	Project Team	Client	Project Office
					TASKS						
	▨		▨	▨ ▨	**41105** Determine contents	▨					▨
▨					**41110** Determine who will work at the project office	▨		▨			▨
▨					**41115** Determine location of contents within the office	▨					▨
▨					**41120** Design the project office	▨					▨
▨					**41125** Assign responsibility for maintenance	▨		▨			▨
					MEASURES OF SUCCESS						
					Are the contents of the office determined?						▨
					Is the office in a central location?						▨
					Is the outlay of the office determined?						▨
					Are the responsibilities for enabling and maintaining the office defined and communicated?						▨
					Is the visibility wall or room meaningful?						▨

7.12 PROJECT MANUAL (41200)

A major step towards greater productivity on a project is developing and maintaining a **project manual**. The manual is a compilation of useful information that project participants can refer to whenever they have a question.

The project manual offers several advantages. It improves communication. If personnel have a reference manual at their disposal, they can find the information they need. If they need to know whom to contact about an approval, for example, they can refer to the manual. They do not have to spend valuable time looking for the information or obtaining incorrect information from someone else.

The manual reduces the occurrence of nonproductive activity. Too often, people spend their time on activities that do not contribute to greater productivity. A common wasteful activity is looking for information that should have been readily available, for example, searching for a phone number or determining the forms to complete in order to purchase or lease equipment. The project manual can help eliminate such wasteful activity by making information readily available.

The manual reduces the incidence of interruptions to productive employees. Someone needing a phone number or a name does not have to interrupt a highly productive individual, as often happens when a new employee interrupts a seasoned project team member.

The project manual offers standardization by containing common information available to everyone. People can all refer to the same procedures and use the same forms rather than perform procedures differently.

Along with standardization comes better control and communication by the project manager. The manual allows you the opportunity to ensure that people have the information to perform effectively and efficiently. It is, above all, a communication tool for you and the project participants.

Finally, a project manual is an excellent tool to train new employees by providing the necessary information to help them to become familiar with the project.

To be useful, a project manual should contain information on many topics, including

- Available resources
- Bar charts
- Change control
- Contingency planning
- Dealing with management and client
- Documentation matrix
- Estimating
- Forms
- Functional priorities
- Meetings (regular and ad hoc)

- Network schedules
- Organization chart
- Phone numbers/listings
- Procedures and policy statements
- Project declaration
- Project procedures
- Reports
- Service support functions
- Statement of work
- Team member responsibilities
- Work breakdown structures
- Work flows

The organization of the manual is just as important as the content. It should not be an assortment of documentation having no logical sequence. If the manual is one large "miscellaneous file," people will set it aside and seek the information elsewhere.

The manual should have an introductory section, which should have a table of contents and present information on how to use the manual. The table of contents contains the reference material and should be arranged by subject.

The manual should also have a section containing supplementary information, such as appendices, glossary, and an index.

Determining the contents and logical construction are two important aspects of developing a project manual. In addition you must address some important administrative considerations.

You need to determine who will plan and organize the manual. Building a project manual is not easy; it requires hard work, time, and money. You want someone who possesses good organizational, research, and writing skills if you want to do a cost-effective job.

You also need to decide who will update the manual during the project. It makes little sense to develop a manual and not keep it up to date. Otherwise, you will eventually have a useless manual. You want your people to refer continually to a manual throughout the project.

The best approach is to select someone who is not working on a critical task. This person will have the time to maintain the manual even if it means letting work on specific activities slide (i.e., consume positive float). The manual should be updated on a regular basis (monthly or quarterly).

Everyone on the project team should have a copy of the manual to increase the availability of the information. If impossible, you can give a copy to certain people, such as team leads or people working on critical path activities. This approach will help lower administrative expenses while ensuring that the right people have access to the right information.

Many project managers shake at the thought of developing a project manual, feeling that it requires too much work and time—and they are right. Yet, the time and effort spent building the manual can later save much more time and effort. Employees will not waste precious time seeking information. By referring to the manual, they will not interrupt already productive employees to ask questions that can be easily answered.

41200 PROJECT MANUAL

P² M² CYCLE — TASKS · RESPONSIBILITIES · INPUTS · OUTPUTS · MEASURES OF SUCCESS

TASKS

Task	Administrative Requirements	Workflows	Updated Reports	Statement of Work	Revised Schedule	Responsibility Matrix	Reports	Project Assignment	Procedures	Organization Chart	Newsletters	New Schedule	New Budget	Network Diagram	Memos	Forms	Contingency Plan	Bar Chart	Project Manager	Senior Management	Project Sponsor	Project Team	Client	Project Manual (OUTPUT)
41205 Assign who will plan and organize the manual	▨																		▨			▨		▨
41210 Assign who will maintain/update the manual	▨																		▨			▨		▨
41215 Determine the outlay/design of the manual	▨																		▨					▨
41220 Determine the contents		▨	▨	▨	▨	▨	▨	▨	▨	▨	▨	▨	▨	▨	▨	▨	▨	▨	▨					▨
41225 Determine the type of binding	▨																		▨					▨
41230 Determine distribution	▨																		▨			▨	▨	▨

MEASURES OF SUCCESS

Measure	Project Manual (OUTPUT)
Are the contents of the manual determined?	▨
Are the contents of the manual complete?	▨
Are the contents of the manual current?	▨
Are the contents of the manual useful?	▨
Are the responsibilities for creating and maintaining the manual defined and communicated?	▨
Is the manual adequately distributed?	▨

8

Project Control (50000)

Introduction

Project plans are not very useful if no one follows them. Successful project managers establish ways to ensure that their projects proceed according to plans.

Status collection and **assessment** are two ways to encourage that project plans are followed. These methods involve collecting data about progress on key activities and other factors and then assessing the impact of performance on the entire project. Project managers often look for and then determine the significance of variances to schedule and budget.

Contingency planning is another way to control projects. Project managers identify possible key scenarios, then develop plans to manage the impact of those scenarios.

Meetings, if conducted well, are a way to control projects. Project managers have three types of meetings at their disposal: status, checkpoint review, and staff.

Change control is a way to bring order out of what might seem like chaos. Most project environments lack complete predictability. When changes arise, the best way to control them is to establish measures for regulating them. Project managers, therefore, establish change control to define, prioritize, and evaluate changes during a project.

Replanning is a way to address change that has rendered the project plan useless. Project managers must replan to develop meaningful, realistic plans to accommodate the change.

An additional control is to take corrective action to get the project back on track. This does not necessarily mean replanning. It does mean taking action that can have both positive and negative impacts. The corrective action ultimately taken depends upon what negative consequences the project manager is willing to allow.

50000 PROJECT CONTROL

P² M² CYCLE

Quality/Design Criteria	Schedule Baseline	Schedule Changes	Budget Changes	Workflows	Quality Changes	Procedures	Schedule Variances	Quality Variances	Contingency Plan	Budget Variances	Major Participants	Management Direction	Potential Schedule Variances	Risk Identification	Potential Quality Variances	Potential Budget Variances	Administrative Requirements	Time Estimates	Defect Rates	Network Diagram	Forms	Cost Estimates	Bar Chart	TASKS
																	■	■	■	■	■	■	■	**50100** Perform status collection and assessment
									■	■	■	■	■	■	■	■	■							**50200** Perform contingency planning
												■					■							**50300** Conduct meetings
■	■	■	■	■	■	■					■	■						■			■	■		**50400** Establish change control
							■	■	■	■	■	■					■							**50500** Perform replanning
											■	■					■							**50600** Take corrective action
																								MEASURES OF SUCCESS
																								Is status collection being performed?
																								Is status assessment being conducted?
																								Are the appropriate meetings being held?
																								Are change control activities occurring?
																								Is replanning occurring?
																								Are corrective actions being taken?

	RESPONSIBILITIES					OUTPUTS																							
	Project Manager	Senior Management	Project Sponsor	Project Team	Client	Impact Analysis	Staff Meetings	Status Review Meetings	Minutes	Checkpoint Review Meetings	Ad Hoc Meetings	Contingency Plan	Action Item Log	Procedures	Updated Schedule	Updated Reports	Updated Budget	Schedule Variances	Quality Variances	Budget Variances	Change Services	Change Board	Change Priorities	Change Categories	Change Baselines	New Quality Requirements	Revised Statement of Work	New Schedule	New Budgets
	X		X		X									X	X	X	X	X	X	X									
	X		X	X	X							X	X	X															
	X			X	X			X	X	X	X	X		X															
	X		X	X		X								X							X	X	X	X	X				
	X		X	X	X									X												X	X	X	X
	X		X	X	X	X												X	X	X									
						X		X	X	X			X	X				X	X	X									
						X				X								X	X	X									
								X	X	X	X	X									X								
						X															X	X	X	X	X				
														X	X	X	X	X	X	X						X	X	X	X
														X								X							

8.1 STATUS COLLECTION AND ASSESSMENT (50100)

When controlling projects, project managers need to perform two overall tasks, tracking and monitoring. Tracking entails reviewing past performance in regard to schedule, budget, and quality. Monitoring entails projecting future results of the project, also in regard to schedule, budget, and quality.

To track and monitor requires performing two basic functions for controlling projects. These are status collection and assessment.

Status collection entails using measures to determine how well certain activities and the entire project are progressing. The most common way is to compile data that will later be converted to information for assessing project performance.

When collecting the status data, you should take the following steps:

1. Collect the status data at set intervals (e.g., every one or two weeks). This will not only ensure consistent data collection procedures, but will encourage people to feel accountable for the progress of their tasks. (The results of any analysis performed will be more reliable and valid based on sound collection procedures.)
2. Collect the status data from all members of the project team.
3. Avoid making judgments when collecting the data. At this time, making premature assessments could affect the reliability and validity analysis of your data.
4. Document the status data you receive. This will encourage people to take their status reporting seriously knowing that they are being held accountable for the data they gave you.
5. Input the data into an automated tool (e.g., a spreadsheet) and generate the appropriate reports.

Assess the project's status by analyzing the project's progress in terms of schedule, budget, and quality. This analysis should include a variance analysis. **Variance** is the difference between planned and actual performance and the formula for its calculation is:

Variance = Plan – Actual

A variance is not necessarily good or bad. It just indicates a difference exists between what was planned and what actually exists; it is simply an indicator to determine if a possible problem exists.

For analyzing **schedule variances**, the project manager should look for the difference between the planned start and end dates and the actual start and end dates for each activity and the entire project. The formula is

Schedule Variance = Planned Start or End Date – Actual Start or End Date

The **budget, or cost, variance** is the difference between budgeted or estimated costs and actual costs incurred for each activity and the entire project. The formula is

Cost Variance = Budgeted Cost – Actual Cost

Once again, the variance indicates an anomaly, but not necessarily something wrong. Project managers need to investigate why the variance exists, ascertain the cause, and determine whether corrective action is required.

50100 STATUS COLLECTION AND ASSESSMENT

P² M² CYCLE — INPUTS · TASKS · RESPONSIBILITIES · OUTPUTS · MEASURES OF SUCCESS

TASKS	Admin. Requirements	Time Estimates	Defect Rates	Network Diagram	Forms	Cost Estimates	Bar Charts	Project Manager	Senior Management	Project Sponsor	Project Team	Client	Procedures	Updated Schedule	Updated Reports	Updated Budget	Schedule Variances	Quality Variances	Budget Variances
50105 Determine what to collect status on		▨		▨	▨	▨	▨	▨						▨	▨	▨	▨	▨	▨
50110 Develop procedure to collect status	▨							▨					▨						
50115 Assign responsibilities for status collection	▨							▨					▨						
50120 Determine what to do with status data	▨							▨		▨		▨	▨						
50125 Determine where to collect status data	▨							▨					▨						
50130 Determine when to collect status data	▨							▨					▨						
50135 Determine medium for collecting status	▨							▨											
50140 Assess schedule performance		▨		▨			▨	▨	▨	▨		▨		▨	▨		▨		
50145 Assess budget performance		▨				▨		▨	▨			▨			▨	▨			▨
50150 Assess performance in regards to quality			▨					▨	▨	▨		▨			▨			▨	

MEASURES OF SUCCESS

	Procedures
Is status collection conducted regularly?	▨
Are the responsibilities for status collection determined and communicated?	▨
Are the appropriate people giving status?	▨
Is the method for collecting status consistent?	▨
Does the method for collecting status lead to reliable results?	▨
Are all meaningful variances identified?	▨
Do assessments involve the appropriate participants?	▨
Are status assessments timely?	▨

8.2 CONTINGENCY PLANNING (50200)

Contingency planning involves anticipating responses to circumstances that can negatively impact a project. This process requires the project manager to determine the necessary steps to overcome anticipated problems.

To a large extent, contingency plans can be made via risk analysis. For each identified risk, project managers provide a description, the probability of occurrence (such as low, medium, or high), the impact (such as low, medium, or high), and an appropriate response or responses. This is often reflected in a completed **contingency plan form** that makes up the contingency plan. **Exhibit 8.2-1** shows an example of a contingency plan form.

Description	Assumptions	Probability	Impact(s)	Response
[1]	[2]	[3]	[4]	[5]

CONTINGENCY PLAN FORM
EXHIBIT 8.2-1

You can use the form in Exhibit 8.2-1 to do contingency planning. Just follow the instructions below:

[1] Identify the problem or circumstance that could arise.
[2] Outline the possible circumstances for which a risk might occur.
[3] Estimate the likelihood of occurrence.
[4] Evaluate how the problem or circumstance will affect the project.
[5] Determine how to deal with the problem or circumstance.

When completing forms, project managers must list all the assumptions influencing their decisions concerning the responses they plan to take. They can later change their responses if the circumstances no longer match the assumptions.

An important consideration for developing contingency plans is obtaining support from all participants affected by a situation. This approval is necessary to enable the project to move forward. Lack of approval is likely to cause the execution of contingency plans to be delayed.

For unforeseen circumstances not captured in a contingency plan, project managers can develop an **action item log**, like the one shown in **Exhibit 8.2-2**. An action item log should contain columns to record the following:

- Comments
- Priority
- Problem description
- Resolution date
  Responsible person

Description	Priority (H, M, L)	Responsible Person	Date of Resolution
[1]	[2]	[3]	[4]

ACTION ITEM LOG
EXHIBIT 8.2-2

You can use the form in Exhibit 8.2-2 to record unforeseen circumstances. Just follow the instructions below:

[1] Create a narrative of the problem or event.
[2] Define the importance of the problem or event.
[3] Identify the individual dealing with the problem or event.
[4] Determine the time when the problem has been or will be resolved.

Project managers can review the action item log during status meetings, which offer an ideal opportunity to add new entries to the action item log.

50200 CONTINGENCY PLANNING

P² M² CYCLE

Administrative Requirements	Major Participants	Management Direction	Potential Schedule Variance	Risk Identification	Potential Quality Variances	Potential Budget Variances	TASKS	Project Manager	Senior Management	Project Sponsor	Project Team	Client	Procedure	Contingency Plan	Action Item Log
		▨		▨			**50205** Develop plans for handling problems in regards to schedule, budget, quality, sponsor, change agents, and change targets	▨		▨	▨			▨	
▨	▨						**50210** Assign responsibilities for contingency planning	▨					▨		
				▨			**50215** Anticipate the repercussions for taking each measure to improve schedule performance	▨		▨				▨	
			▨	▨			**50220** Determine the measures to take to deal with deviations to the schedule	▨						▨	
				▨		▨	**50225** Determine the measures to take to deal with deviations to the budget plans	▨						▨	
				▨	▨		**50230** Determine the measures to take to deal with deviations to the plans regarding quality	▨						▨	
▨	▨	▨					**50235** Develop a procedure for performing contingency planning	▨					▨		
	▨			▨			**50240** Conduct impact analysis for potential problems	▨		▨				▨	
				▨			**50245** Anticipate the repercussions for taking each measure to improve budget performance	▨						▨	
				▨			**50250** Anticipate the repercussions for taking each measure to improve quality of performance	▨						▨	
▨	▨						**50255** Log and track unforeseen issues	▨		▨	▨	▨			▨
							MEASURES OF SUCCESS								
							Are contingency plans based upon risk control results?							▨	
							Are all the right participants involved in contingency planning?						▨		
							Are contingency plans in sufficient detail?							▨	
							Are contingency plans documented and communicated?						▨	▨	
							Is a mechanism in place to log and track unforeseen issues?								▨

8.3 MEETINGS (50300)

Meetings are group sessions that should result in accomplishing goals and objectives efficiently and effectively.

Unfortunately, meetings are not always efficient and effective. Quite often, they are too long, unfocused, dominated by a few individuals, and provide insufficient documentation of the results.

To have meaningful meetings, adhere to the following important rules:

1. Announce the meeting far enough in advance to allow everyone sufficient time to prepare for it and to arrange their schedules.
2. Prepare an agenda and keep to it during the meeting. An agenda will help prevent the tendency to go off on a tangent and, consequently, not address the main issues. You should distribute the agenda in advance so that everyone can be prepared.
3. Document the results of the meeting. This documenting does not necessarily mean writing down every comment but should include the decisions made at the meeting and by whom. A summary document will breed greater accountability and commitment on the part of attendees and will serve as an audit trail for resolving project problems.
4. Invite all the relevant people. A good meeting encourages participation by everyone who can add value to the discussion. Failure to invite all relevant parties can result in creating suspicion among team members and other project participants and result in a lack of commitment.
5. Encourage everyone to participate during the meetings. Avoid letting a few people dominate the conversation. Otherwise, resentment can arise, commitment can wane, and poor feedback can result.
6. Avoid long meetings. People's attention span and biological needs will only compete with the purpose of the meeting. If a meeting figures to last more than one hour, consider scheduling periodic breaks to enable participants to re-focus on the purpose.

There are two basic kinds of project meetings: planned and ad hoc. Planned meetings occur at set intervals or designated points during the course of a project. There are three types: checkpoint review, status review, and staff.

Checkpoint review meetings occur whenever a project finishes a certain phase, task, or milestone. Key team members get together to discuss progress to-date and may even decide whether or not to proceed.

Status review meetings are for collecting and assessing status on a regular basis to ascertain progress. A key difference from the checkpoint review meeting is that members do not need to decide whether or not to proceed.

The **staff meeting** is a weekly or biweekly meeting held for team members to deliver announcements or exchange information concerning the project.

Ad hoc meetings are held whenever a need arises. Don't be fooled by the spontaneity of these meetings. Documenting their results and having the right participants are just as critical as for planned meetings. Hence, being ad hoc does not necessarily mean being informal.

50300 MEETINGS

		P² M² CYCLE / TASKS	Project Manager	Senior Management	Project Sponsor	Project Team	Client	Procedures	Status Review Meeting	Staff Meeting	Minutes	Checkpoint Review Meeting	Ad Hoc Meetings
INPUTS			**RESPONSIBILITIES**					**OUTPUTS**					
Administrative Requirements	Management Direction	**TASKS**											
	▩	**50305** Determine types of meetings required	▩					▩	▩	▩		▩	▩
▩		**50310** Assign responsibilities for meetings	▩		▩			▩	▩	▩	▩	▩	▩
▩		**50315** Determine medium for minutes	▩								▩		
▩		**50320** Determine location	▩			▩	▩		▩	▩		▩	▩
▩		**50325** Determine when to hold	▩			▩	▩		▩	▩		▩	▩
▩		**50330** Develop an agenda	▩			▩	▩		▩	▩		▩	
▩	▩	**50335** Designate attendees	▩			▩	▩	▩	▩	▩		▩	▩
▩		**50340** Record minutes	▩		▩			▩	▩	▩	▩	▩	▩
		MEASURES OF SUCCESS											
		Are the right types of meetings determined?							▩	▩		▩	▩
		Are the right participants determined for each type of meeting?						▩					
		Does each type of meeting have a standard format to follow?						▩	▩	▩	▩	▩	▩
		Are responsibilities defined and communicated for each type of meeting?						▩					

8.4 CHANGE CONTROL (50400)

Projects are easy to manage if everything is static—if no changes occur and everything moves along according to plan. Yet, reality proves otherwise: nothing is ever perfect.

This tendency towards change requires implementing measures to prevent the project manager from falling into a reactive style of management. Change control is one way to prevent being inundated by constant change.

Change control involves policies and procedures to detect, analyze, evaluate, and implement modifications to all baselines in your project. A baseline is an agreement between two or more parties on what constitutes something, such as a product description, schedule, budget, and quality levels. Any changes to these baselines may severely impact the project. It behooves project managers to control changes so that changes do not control them.

Sources of Change

Project managers are in a delicate position because they interact with many people and organizations related to the project, directly or indirectly. These people or organizations make decisions requiring a change; project managers must respond to these circumstances. The agents initiating change include

- Customer
- External authority (i.e., government)
- Project manager
- Project team
- Project sponsor

This list is not exhaustive. Changes can come from sources beyond any person's control. You may face technological changes, such as to your product baseline, to keep pace with the state of the art. Or you may confront an unexpected downturn in your business or the economy, thereby causing you to reduce the size or scope of your project. Or an "act of God," such as a flood of the office, may cause the schedule to slide. Or an "act of man," such as an unexpected labor strike, may also require changing all baselines.

As a project manager, you must identify changes and understand their impacts on your project. These impacts often affect many areas.

Impact of Change

Change can be far-reaching and can impact several areas of your project. Change can affect people by altering their feelings about themselves, their job, and their company. A negative reaction can result in absenteeism and turnover, leading to a decline in productivity.

People do not fear change itself, only the way they are introduced to it. Random and chaotic changes can make people defensive, even aggressive. Change can also affect the way people do business on a project. If employees are accustomed to doing business a specific way and someone introduces a change, they can become upset or frustrated.

Change can affect the schedule. A change to the product baseline may extend or shorten the project completion date. Further, it can affect the budget directly or indirectly. A change in the schedule can mean reducing or enlarging the budget. Or funding allocated to the project could be directly cut by a certain percentage by management or the customer.

Change may affect the level of quality of work. If management has shifted its priorities and feels you or your team should devote less energy to the project, you and your team may be able to concentrate on only the more important areas.

Finally, change may affect the final product and may mean alternating its appearance and functionality. For instance, a change may impact hardware, software, or documentation (e.g., user manuals or reports).

You need to track and manage all changes. That may sound like common sense, but the reality is that many project managers fail to do so. Instead, they react to changes by treating each one as equally important instead of prioritizing them.

Categories of Change

As a project manager, you will receive many changes from many sources impacting many areas. You can quickly become inundated, even overwhelmed, by the changes unless you systematically handle them.

Categorizing your changes will allow you to group them logically according to their expected impact. You can always re-categorize a change based upon a more detailed analysis later.

You can divide changes into four categories.

1. A **major change** dramatically alters the schedule, the product baseline, the budget, and/or anything else deemed critical to the project. A major change dramatically alters the outcome of the project. Two examples a major change are management dictums to reduce or increase the level of functionality of the final product and cutting the project budget.

2. A **minor change** does not imply insignificance. It is categorized as minor only because it fails to fit in the major change category. A minor change does not alter the ultimate outcome of the project. It does, however, potentially affect the chances for a successful outcome.

3. A **corrective change** fixes something that was overlooked sometime during the project. For instance, if a section of a marketing proposal was omitted, you would go back and include it.

4. A **maintenance change** is required to keep something current. It has very minimal impact on the outcome of the project or the current mode of operation. It affects something that is already completed but that needs to be updated.

Priority of Changes

After categorizing the changes, you assign a priority to each one. You classify a change as major priority, minor priority, or no priority.

A change classified as a **major priority** needs attention immediately, otherwise, it will be a "show stopper." Not addressing it immediately could jeopardize the project's schedule, budget, or quality.

A change that can be considered a **minor priority** does not require immediate attention, but you must confront it eventually to complete the project. In other words, you can continue for a while as planned, but sooner or later you must incorporate the change.

A change having **no priority** may be addressed if the time permits. You do not have to incorporate the change and can still finish the project successfully.

For changes designated as a major or minor priority, you should assign a date when they will be implemented. This date is called a block-point date.

Change Board

Project managers do not classify or assign a priority themselves to a change in large projects. Typically, they, along with several select others, form a **change board**.

The change board comprises the project manager, team leaders, project sponsor, and customer representatives. Meeting at regular intervals, they

- Analyze the impact of changes
- Assign priorities to changes
- Assign responsibilities for implementing changes
- Classify, or categorize, incoming changes
- Decide the fate of changes (approve or disapprove)

Project managers have the ultimate responsibility for the decisions of the change board. Depending on their managerial style, project managers may make decisions themselves and treat the board as a group of advisors, require a majority vote on decisions, or require a unanimous vote.

Change Control Procedure

Whether your project is small, medium, or large, you should have a procedure for managing changes. For small projects, the procedure can be very informal. For medium or large ones, however, you will need formal change control procedures, because the number of people involved on the project increases. The best way to formalize change control is to develop written documents.

Your **change control procedure** should describe the steps from the moment the need for a change arises to its implementation. It should address who, what, where, when, why, and how.

Exhibit 8.4-1 is a work flow describing a typical change control process. You may elect to prepare separate procedures for each step or one describing the entire process. Generally, the more detail the better, as long as the procedures do not stifle productivity.

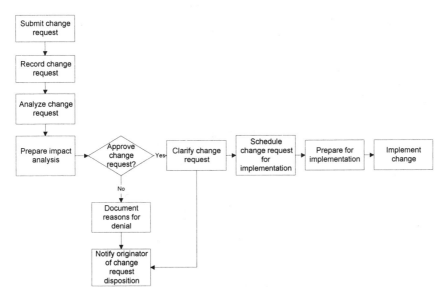

TYPICAL CHANGE CONTROL PROCESS
EXHIBIT 8.4-1

You will notice, too, that the work flow begins with submitting a change request. An illustration of a change request is shown in **Exhibit 7.6-2** (p. 179).

Having a form like the one shown in **Exhibit 8.4-2** serves several purposes. First, it functions as a historical record of changes. Second, it instills a sense of accountability on the project by recording the judgments of decision makers. Third, it forces participants to think analytically because it raises issues concerning the impact of a change.

You can use the form in **Exhibit 8.4-2** as a change request for your project. Just follow the instructions in **Exhibit 8.4-2**.

Date	Change Description	Impact Analysis	Priority	Initiator	Person Responsible	Change Board Approval?	Effective Date
[1]	[2]	[3]	[4]	[5]	[6]	[7]	[8]

[1] Calendar date for entry into the log
[2] A narrative description of what the change is
[3] The consequences of the change, such as the schedule, quality, and budget
[4] The importance of the change, such as high, medium, or low
[5] The person or organization submitting the change
[6] The person or organization to implement the change
[7] A "yes" or "no" indicating whether the change received approval by the change board
[8] The date when the change will go into effect

CHANGE CONTROL LOG
EXHIBIT 8.4-2

Procedures for the operation of a change board should delineate the who's, what's, when's, where's, why's, and how's of the entire change board. The following is a possible outline for this procedure:

    I. Introduction
        A. Purpose
        B. Scope
   II. When the board meets
        A. Date
        B. Time
  III. Where the board meets
        A. Location
        B. Necessary facilities
  IV. Who participates
        A. Attendees
        B. Responsibilities of attendees
        C. What constitutes a quorum
    V. How business should occur
        A. Conduct of proceedings
        B. Disposition of change requests
        C. Use of documentation
  VI. What topics the board addresses
        A. Old business topics
        B. New business topics

Conclusion

Projects do not occur on an isolated island. Their environment is dynamic, not static, and they interact with a wide variety of variables. A change can easily upset the equilibrium of projects. It behooves project managers to control change before it controls them. If the latter occurs, project managers can quickly become reactive.

50400 CHANGE CONTROL

P² M² CYCLE

TASKS

- 50405 Develop formal change control procedures
- 50410 Identify categories of change
- 50415 Identify priorities of change
- 50420 Identify sources of change
- 50425 Establish baselines
- 50430 Establish a change board
- 50435 Define change board membership
- 50440 Define change board functions
- 50445 Perform impact analysis

MEASURES OF SUCCESS

- Is a change control procedure in place?
- Have the different categories of changes been identified?
- Have the different priorities of changes been defined?
- Are the membership and responsibilities of a change board defined?
- Does a change board exist?
- Are impact analyses performed?
- Has a change board person been designated?
- Are change board members identified?
- Is a change board regularly scheduled?
- Are submitters completing the change request form?
- Are the changes being implemented as expected?
- Is someone filing the change board meeting minutes and documentation?

INPUTS

Task	Inputs
50405	Forms, Procedures, Workflows
50410	Quality Changes, Budget Changes, Schedule Changes
50415	Procedures, Quality Changes, Budget Changes, Schedule Changes
50420	Cost Estimates, Quality Changes, Time Estimates, Schedule Baseline, Quality/Design Criteria
50425	Procedures, Quality Changes, Major Participants
50430	Procedures, Major Participants, Management Direction
50435	Procedures, Major Participants, Management Direction
50440	Procedures, Management Direction
50445	Cost Estimates, Quality Changes, Time Estimates, Budget Changes, Schedule Changes, Schedule Baseline, Quality/Design Criteria, Management Direction

RESPONSIBILITIES

Task	Responsibilities
50405	Project Manager
50410	Project Manager
50415	Project Manager
50420	Project Manager
50425	Project Manager
50430	Project Manager, Project Team, Client
50435	Project Manager, Project Team, Client
50440	Project Manager
50445	Project Manager, Project Team, Client

OUTPUTS

Task	Outputs
50405	Procedures
50410	Change Categories
50415	Change Priorities
50420	Change Sources
50425	Change Baseline
50430	Change Board, Procedures
50435	Change Board
50440	Change Board, Procedures
50445	Impact Analysis

8.5 REPLANNING (50500)

Under some circumstances, the options for corrective action will not work. You will have no other choice but to replan. Replanning requires restructuring part or all of your project. Before deciding to replan, consider these options.

Replanning takes time—sometimes a great amount of it. Although one can hope that the time spent for replanning will result in more time for production, the fact remains that replanning consumes time ordinarily devoted to production. Besides time, it also takes people, equipment, and supplies to replan effectively. You must divert these resources from the production line.

Change is another consequence of replanning. In trying to institute change, you will face some resistance. Some team members may have liked the initial schedule and the budget allocated for each activity. They may not want replanning to now change everything and may even view replanning as threatening, not appreciating that your purpose is to improve productivity and manage a project to finish on time, within budget, and at a satisfactory level of quality.

You must solicit participation for replanning from all the major participants in the project. You need the support of the customer as far as making changes in major milestone dates and budgeting. You will need the support of your team as far as providing feedback and buy-in to the new plan. And you will need the support of your management, as they will ultimately decide whether to allocate more funds or other resources to the replanning effort.

Consequently, you want to obtain approval from everyone involved in the replanning effort. You should obtain their signatures or initials on every major item that is changing (e.g., schedules, deliverables, and work breakdown structure). Once everyone from the client, project team, project sponsor, and senior management side agrees, they will find it difficult to criticize and not abide by the new plan.

Because replanning involves considerable time and effort, be aware that you may have to change several items used to manage the project. You may have to change, for example, one or more of these items:

- Administrative procedures
- Approvals
- Budgets
- Estimates
- Goals and objectives
- Levels of quality
- Managerial style
- Methods of communication
- Network diagram
- Organizational chart

- Priorities
- Project manual
- Responsibility matrix
- Statement of work
- Technical documentation
- Types and numbers of meetings
- Types and quantities of forms
- Types and quantities of reports
- Utilization of supplies
- Utilization of nonlabor resources
- Work breakdown structure
- Work flow of operations

How to Replan

If electing to replan, do so in a manner that encourages greater productivity and does not impede it. Replanning requires being methodical, not whimsical. The goal is to replan without drastically impacting productivity to the point that it causes schedules to slide, increases budget overruns, or lowers quality. To replan effectively, consider these steps:

1. Identify the reasons for the replanning effort. Locate the source of the problem and fix it. Avoid replanning to fix a symptom and not the cause, which simply leaves the necessity for another replanning effort. Chronic replanning does not look good for project managers and only contributes to schedule slides, budget overruns, and lower quality.

2. Identify exactly what you must change by conducting an impact analysis. You should answer questions like: Will I have to change the statement of work? The work breakdown structure? The network diagram? Responsibility assignments? The budget? Ways of doing business? The demographics? Our suppliers? You should answer these and any other questions you have before replanning gains momentum. After all, you should know what you are trying to achieve.

3. Determine how long the replanning effort will take. If replanning is taking too long, it can contribute to an even greater schedule slide. If too short, it can prove a waste of time if it lacks thoroughness of detail.

4. Determine who should participate in the replanning effort. A good guideline is to invite individuals affected by the effort. Avoid omitting people from the replanning effort, because they will tend to either sabotage or fail to acknowledge the validity of your new plans.

5. Determine the resources and their quantity required to replan. Determine the number of people with the appropriate skills. Select the right equip-

ment and facilities to replan. Calculate and make available the necessary amount of time and money to replan cost-effectively.

6. Hold individual sessions with participants to obtain their participation in the replanning effort. Ask them questions about how to make the plan better. Using sessions with individuals to acquire the information helps you to discuss matters more frankly and collect the necessary data.

7. Select data and suggestions from the individual sessions and include them in your new plans. Be aware that changes may have to affect the statement of work, work breakdown structure, or responsibility assignments.

8. Hold group sessions to resolve any differences of opinion among participants. Having everyone present at the meeting will allow for building a consensus. You might find it best to hold the group meetings away from your work site to provide an opportunity to build esprit de corps and avoid having the team collapse into factions. You will find it important to get everyone's agreement at the outset rather than after new plans are implemented.

9. Finalize any last minute revisions. Acquire everyone's signatures on the appropriate documents. Provide copies of the new plans to each new project participant.

Using Good Change Management

Replanning involves change, which does not come easy to many people. Once a project gains momentum, changing its course becomes difficult; you often face resistance. **Change management** can help you overcome this resistance.

The goal of change management is to introduce change in a manner that minimizes negative impacts on productivity.

One of the first steps in change management is to prepare people for change. This means not giving them surprises or catching them off guard. If you fail to prepare them, you will be much more likely to face resistance. Turnover and absenteeism are two ways people express their disapproval of change. Other ways include sabotaging change (such as not complying with it) and finding alternatives to change.

Several ways exists to prepare people for change. Your should first strive to acquire the participation of everyone impacted by change. If you change the work breakdown structure, for example, determine who will be impacted by the change and solicit their feedback. Also, require their signature on all applicable documentation to record their approval for changes in estimates, to the statement of work, and to other important documentation. By acquiring their participation, you expose any potential opposition.

Also, strive for open communications up and down the chain of command and laterally with the customer. Be up-front with all parties about any changes. The minute suspicions arise over your motives and credibility, you will face little support for your new plans.

In addition, prepare everyone for the new plans. Avoid surprises to preclude resistance. Hold "mini" training sessions about the new plans, highlighting what has and has not changed and why. Give people new documentation to use. Let them know whom to contact for questions. Remove as many obstacles to the change as possible before it takes effect. In summary, prepare prior to implementing a change.

You must sell the change. People will accept change if they realize its benefits and the problems or costs with maintaining the status quo. Effective selling makes people think they need a change. You need to explain, from their perspective, why the change is necessary to the project team. You should avoid giving them the impression that change will mean instant improvement and will go smoothly. You should mention that difficulties will arise and provide suggestions to overcome them.

As a project manager, you need patience to deal with change, because handling its dynamics is strenuous. Not everything will go according to your new plan; indeed, your new plans may occasionally need revisions. You need to psychologically adapt to handle the stress accompanied by change.

You can use peer pressure to institute change. Occasionally, project participants resist change for the sake of it. They will do anything to obstruct change. Using peer pressure can often turn the most obstinate mind around. If others support the change, for example, have them sell for you. Let them pressure your opposition to support your changes. Peer pressure is a powerful weapon to urge people to comply with your wishes. If opposition continues, do what most people do when facing a brick wall; either go around it or tear it down.

You must also recognize that change requires three participating parties: your team, your management, and the customer. Without their support, you will find implementing change very difficult. If you are changing the schedule without the customer's concurrence, for example, you will probably lack cooperation and may face litigation. If you fail to consult with your project team, you may find that certain members will feel like "pawns" and become alienated and, consequently, will perform at a reduced level. If you failed to consult with upper management, they may feel slighted and respond by decreasing future political and financial support.

Finally, remember that good change management requires playing politics. A change may make good rational, technical sense. But this alone may not be enough to implement a change effectively. You must recognize that effective change involves role playing. You have three roles to deal with change management: change target, change sponsor, and change agent.

The **change target**, which may be a person, place, organization, or thing, is the object of the change. Change targets respond to change to varying degrees. A change target could be the customer or the project team. In some instances, it could be senior management.

The **change sponsor** may be a person (or persons) or an organization. It provides the political "muscle" to effect a change. The sponsor can assist by announcing its support to the change. Typically, change sponsors come from senior management.

The **change agent** may be a person, a group, or an organization. It actually plans, organizes, and executes a change, working as a catalyst for a change. Project managers, because of their unique position, are usually change agents.

Conclusion

Managing a project is not easy. You can build the best plans in the world and still fail. Quite often, the failure results from a failure to manage the project according to the original plan. Effective project managers must know when activities are straying from the plan. If you notice that your project is straying from your plan, take the necessary action to regain control. If you find your plan is no longer relevant, replan.

50500 REPLANNING

P² M² CYCLE — INPUTS · TASKS · RESPONSIBILITIES · OUTPUTS · MEASURES OF SUCCESS

INPUTS: Major Participants	Administrative Responsibilities	Schedule Variances	Quality Variances	Management Direction	Contingency Plan	Budget Variances	TASKS	Project Manager	Senior Management	Project Sponsor	Project Team	Client	New Quality Requirements	Procedures	Revised Statement of Work	New Schedule	New Budget
▨	▨			▨			**50505** Establish a procedure for replanning	▨						▨			
▨	▨				▨		**50510** Assign responsibilities for replanning	▨			▨			▨			
	▨						**50515** Determine resources and quantities required	▨						▨			
		▨	▨			▨	**50520** Determine the circumstances under which to replan	▨						▨			
	▨						**50525** Determine the resources needed for replanning	▨						▨			
	▨						**50530** Identify participants	▨			▨	▨		▨			
	▨	▨	▨	▨			**50535** Determine duration of replanning	▨			▨	▨		▨			
▨						▨	**50540** Identify reasons for replanning effort	▨		▨				▨			
							50545 AObtain participation from all affected participants of the replanning	▨			▨		▨				
		▨	▨			▨	**50550** Perform an impact analysis	▨							▨	▨	▨
							MEASURES OF SUCCESS										
							Does a procedure exist for replanning?							▨			
							Have the responsibilities for replanning been determined and communicated?							▨			
							Are all the participants in replanning activities identified?							▨			
							Is an impact analysis conducted for replanning activities?						▨		▨	▨	▨

8.6 CORRECTIVE ACTION (50600)

Project managers must frequently make an existential decision. They must decide whether to follow their plans or discard them and, if the latter, how to develop and implement new ones. These decisions reflect their style of management.

Whether to Follow Your Plans

Reactive project managers stray from their plans at the earliest opportunity. They develop plans, often meticulous ones, to satisfy other parties such as the customer or senior management. Once their project begins, however, they set aside their plans and start replanning or quit planning. They begin reacting, which leads to cost overruns, schedule slides, and quality degradation.

Proactive project managers attempt to follow their plans as much as possible. When their project strays from their plans, they take necessary actions to synchronize their circumstances and plans. Only if all attempts fail do they replan.

Proactive project managers act in a way that is flexible but not fluid. The distinction is important. Flexible means that they are not averse to change but that they adopt it only when necessary.

Fluid project managers change for the sake of changing. The minute their project goes awry, they quickly deviate from their original plans. Irrelevant plans and constant replanning characterize fluid project managers.

The key to effective control of projects is, therefore, to strive to follow your plans. If straying from your plans, you should attempt to take the necessary action to synchronize your plan to the circumstances.

Many project managers think that they must discard their plans whenever they face schedule slides, budget overruns, or quality problems. They often do not realize that there are several corrective they can take before discarding their existing plans.

Getting Back on Schedule

If your project falls behind schedule, you can pursue several measures. However, each measure has one or more side effects.

Rescheduling is one measure that you can take. After careful analysis, you may find gross inaccuracies in the estimates and dependencies in the schedule. Or, you may find that the assigned responsibilities and assignments are unrealistic. These and other shortcomings could lead to a major rescheduling effort.

If you elect rescheduling, you will face certain consequences. Rescheduling requires that time be removed from actual production. It also delays the project for a short while; eventually, rescheduling will lead to a project that is on schedule. Finally, rescheduling requires change, which can frustrate project participants, who may resent changes to the current mode of operation and may need to overcome a learning curve.

Acquiring more competent people can help improve compliance with your schedule. Team members with greater expertise can increase productivity by working faster and smarter than the original team members.

Drawbacks exist to acquiring more competent employees. If acquiring them late in the project, their learning curve, no matter how high their competence level, could slow progress, not increase it. The new participants may fail to acquire the familiarity necessary to become fully productive.

This approach can also lead to declining morale of existing team members. Original members may feel threatened by highly competent newcomers. They may perceive, rightly or wrongly, that the new members will rob them of opportunities for growth.

Finally, this approach can result in higher labor costs. People with greater expertise will demand more compensation than those already working on the project.

Purchasing or leasing better equipment is another way to improve your schedule status. Using more powerful computer software and hardware, if introduced correctly, may increase productivity.

Purchasing or leasing better equipment does have drawbacks. It not only increases costs, but can slow productivity if people lack training or feel hostile towards it. New equipment requires change and; if not managed correctly, its introduction can result in morale problems.

Streamlining operations is another measure to improve your circumstances. Perhaps team members require too many approvals to do anything. Or they spend more time completing forms than producing. Removing these and other obstacles can enable you to get your project back on schedule.

Streamlining operations requires diverting valuable time from productive work. Your team will need to identify what needs improvement and to learn new ways of operating, which can lower productivity in the short run but hopefully increase it in the long run.

Another way to improve on your schedule performance is to let the team lower its performance standards (quality). This measure will improve schedule performance but can have very serious consequences.

Lower quality sets a bad precedent for future performance. It signals to team members that lower quality is permissible. Lower quality can result in a bad reputation for yourself or your team, or both. A loss in reputation can mean less credibility and an eventual decline in opportunities for future projects and can even result in legal complications.

Employing phased implementation is another measure to pursue. You can deliver your final product to the customer in batches. If developing a major report, you may elect to release specific parts to the customer. Or you may elect to release a scaled-down version of the final product and upgrade it to a more sophisticated version later.

This approach, too, has repercussions. The customer can become disappointed over the delivery of the interim product. The approach can also result in protracting the schedule, a problem that is common on large software projects. Typically, the customer initially receives a smaller system than originally requested and later on receives enhanced versions.

Concentrating only on the critical path activities in your network diagram is another means for improving schedule performance. You devote all your energies to those particular activities until they are finished, which improves the chances that the project will be on schedule.

Negative consequences do exist even when pursuing this approach. Activities slide, often to the point that they become part of the critical path. The momentum for completing the project slows down and may be difficult to regain later.

Use of overtime work is a popular method project managers use to improve schedule performance. But it is best regarded as a short-term solution to compensate for lost time.

Overtime, especially when excessive and unplanned, can burn out staff, causing a severe decline in morale and contributing to high turnover. It also can lead to lower productivity. Some people produce less during normal work hours and use overtime to finish their workload and invent an opportunity to increase their income. Other employees, after extensive overtime, make serious production errors because of declining attention span.

Excessive overtime does not reflect well on project managers. They appear highly reactive in their approach and not in control. They fail to use their labor pool cost-effectively, and they appear to lack the wherewithal to manage the project.

Shift work is an alternative to overtime. People work back-to-back, usually during daylight and evening hours. If used correctly, shift work gives you the opportunity to meet the milestones in your schedule.

But shift work, like other techniques, has its drawbacks. You must coordinate production among both shifts so that duplication of effort does not occur. This requires ensuring communication between shifts to prevent misinterpretations. Finally, shift work can increase project costs if shift differential rates are applied.

Terminating vacations and training can help schedule performance by making available personnel who would have been on vacation or in training.

But terminating vacations and training can lower morale and, consequently, productivity. In the short run, you may gain time and improve schedule performance; in the long run, you can face lower productivity and more schedule slides.

Using negative and positive incentives is another option. Sometimes the "stick" will encourage people to produce more and even better. At other times, the "carrot" will entice people to work harder and more effectively.

Negative incentives work only in the short run. Although negative incentives, such as counseling or letters of reprimand, may improve schedule performance temporarily, resentment by employees can remain.

Positive incentives can lead to the constant expectation that "I will perform well only if given something more than my pay." This may lead to difficulty in motivating people when the positive incentives are no longer available.

Upgrading employee skill levels is another way to improve schedule performance. Sending people to training is one way to accomplish that. You might elect to send employees to seminars or institute an on-the-job training program. This can "jump-start" what they need to do to deliver.

Of course, it must be remembered that training sessions require time and that people are not producing while they are attending them. Another negative is the cost; training can be expensive.

Improving communications is another option to employ. Your project may be falling behind schedule because people lack the information to perform their tasks. You may need to hold more meaningful meetings and improve the accessibility and distribution of project documentation.

The negative consequences are that people attending meetings are not directly producing. Or they may already attend too many meetings (but not meaningful ones) and receive too much unnecessary documentation, thereby leading to greater schedule slides.

Reducing Costs

You may occasionally face budget overruns, forcing you to spend more money than you had originally expected. As a project manager, you can exercise options to reduce your current expenditures.

Downsizing the scope of your project is one measure to employ. You deliver to your customer a scaled-down version of the product. One possible negative consequence is obvious: the customer becomes disappointed because the product does not fulfill its original expectations.

Reducing the size of your staff is another option. For instance, you can release people who are not working on the activities on the critical path. Such an approach has some negative impacts. You may lose valuable, highly skilled people even though they are not working on activities on the critical path; you may never get those people back. You may also eventually have to acquire new people to replace them; and these people must overcome a learning curve before becoming fully productive.

Hiring cheaper labor is a technique to lower project costs. Tapping a cheaper labor pool will lower not only current costs but future ones, too. But using cheaper labor has a significant risk. Your labor may be cheaper but not necessarily more competent. The quality of work may decline to a degree that the money saved may not exceed the loss in poor performance. In other words, "you get what you pay for." You may face another negative impact in that your team members may see cheaper labor as a threat—a means of replacing them. The morale of the team can rapidly decline along with productivity.

Using cheaper equipment, materials, and supplies will reduce project costs, too. After careful shopping, you may, for example, find that team members can save on hourly or daily leasing rates by using automated machinery. Less expensive materials and supplies, including quantity discounts, also saves money.

Naturally, using cheaper equipment, materials, and supplies can have negative consequences. Cheaper equipment, if it is inferior, may result in a slower production rate, which in turn can contribute to schedule slides. Using cheaper materials and supplies may result in lowering the quality of the final product.

Cutting training can reduce costs. You simply reduce the amount of training available to team members. For instance, you might send someone to a two-day project management seminar rather than a five-day one.

By reducing training, however, you may incur the wrath of some team members. Many see training as a reward for outstanding performance or an opportunity for career growth. Any attempt to curtail training can cause people to become upset and reduce productivity. Furthermore, you may increase project costs because some members will not improve their skills, which can result in lower project costs.

Reducing overtime is another effective technique to decrease project costs. You can eliminate overtime completely or restrict the amount, such as no more than five hours per week. Since overtime is usually charged at one-and-a-half or twice the normal hourly rate, restricting its use can help keep the labor costs of the project in check.

Cutting overtime can have negative outcomes. Team members may see overtime as akin to a pay raise and may have come to expect the extra income. After reducing overtime, you are, in effect, reducing that extra income.

Streamlining operations can reduce project costs. Decreasing the number of meetings, reducing the number of approvals to complete tasks, and restricting the amount of forms to complete are ways to lower costs.

This technique, like all other ones, has its negative consequences. People need time to learn the new mode of operation. People may become frustrated, having to learn and adjust to a new procedures.

Finally, improving communications can reduce costs. You can establish mechanisms that ensure team members receive the information to preclude duplicated effort and needless reiteration.

Ironically, improving communications may lead to problems. You and team members must take time away from productive work to establish good communications within your project. Also, you must provide time for people to learn how your new communication structure works.

Improving Quality

In addition to improving performance regarding schedule and budget, sometimes you must improve quality.

Quality means something different to everyone. In this book, quality means developing a product that complies with the requirements of the customer and the industry. Several ways exist to ensure that the product satisfies the requirements and desires of the customer and the standards established by the industry.

Increasing performance standards can improve quality. You might lower the tolerance level, for instance, on the number of defects in the product. To enforce compliance with the new standard, you would need to track the number of defects and who was responsible for them.

Instituting vigorous change control can also improve quality. You can establish procedures for managing changes to ensure that the customer receives the product requested. When changes to the requirements occur, you establish procedures to track them, assess their impact, prioritize them, and manage their disposition.

Hiring highly competent and highly skilled people increases the potential for better quality. However, hiring people merely according to good credentials does not guarantee higher quality. These people must have the motivation to perform well. Some project managers mistakenly believe that hiring more qualified people will guarantee an improvement in quality.

Concentrating on only the important aspects of the project will improve quality. Not all tasks are critical in terms of the functionality of the product. You want to identify those tasks that will contribute most to the quality of the product. You must devote the majority of your team's time and resources to activities important for building a product of the highest quality.

Inviting audits of the technical features of your product can effectively improve quality. If your organization is large, you can ask someone from another department to evaluate the quality of work by your team. If not, you can hire an outside consultant.

The advantage to having someone not on the project team evaluate quality is obvious. When someone is not emotionally involved with the output, defects can be more easily detected. The auditors can also make recommendations for improvement without fear of reprisal. You can start conducting quality reviews. Employees hold the reviews whenever a significant part of the final product or a major phase is complete. Team members congregate in a conference room to assess the quality of output. If they deem the output meets standards, they approve it. If not, they recommend improvements and do whatever is required to meet the standards.

Encouraging cross-checking of work can improve quality. After a team member completes a task that results in some tangible output, a colleague reviews it. Cross-checking allows for detection of errors often overlooked by the original developer. Unlike quality reviews, cross-checking is less formal and involves fewer people.

Training employees can increase the quality. Whether through on-the-job, classroom, or self-study training, team members can enhance the quality of their own output. Armed with greater expertise, they will perform faster and make fewer errors.

Acquiring better equipment and supplies can improve quality. You could, for example, purchase more expensive materials to build a house to prolong its life. Likewise, better equipment can save time and allow team members to concentrate on quality rather than quantity.

Using negative and positive incentives can improve quality. You can use negative measures, such as counseling sessions and letters of reprimand, to spur people to improve their quality of work.

In addition, you can use positive incentives, such as bonuses and compensatory time, to encourage people to concentrate on the quality of their work. You might consider collecting performance data on each team member and reward the individual with the highest volume of output with the least number of defects.

Finally, improving communications can increase quality. Perhaps people are not performing high-quality work precisely because they lack the information to do their job correctly. This circumstance occurs quite frequently in data processing. People building the computer system lack all the specifications to develop software of superior quality.

Although you have several options available for improving quality, exercising them can have unforeseen repercussions.

Instituting quality control measures can slow the project. Encouraging cross-checking can breed competition and foster a negative attitude among team members. Inviting audits can undermine your own managerial authority, especially if auditors discover something serious and inform high-level management. Hiring highly competent (i.e., more highly paid) people can shoot project costs upward and disrupt existing team members. Instituting rigorous change control can slow schedule performance and add another level of bureaucracy.

Whether dealing with schedule, budget, or quality, you have options for improving circumstances. Ironically, sometimes exercising these options can result in magnifying your problem. In all cases, you must weigh carefully which options to select.

50600 CORRECTIVE ACTION

INPUTS						P² M² CYCLE	RESPONSIBILITIES					OUTPUTS				
Management Direction	Administrative Requirements	Budget Variances	Quality Variances	Schedule Variances	Major Participants		Project Manager	Senior Management	Project Sponsor	Project Team	Client	New Quality Criteria	Impact Analysis	Procedures	New Schedule	New Budget
						TASKS										
					▩	**50605** Establish a procedure for taking corrective action	▩							▩		
					▩	**50610** Obtain participation from all affected participants of the corrective action	▩		▩	▩	▩	▩			▩	▩
		▩	▩	▩		**50615** Identify reasons for corrective action	▩						▩			
▩	▩					**50620** Determine duration of corrective action	▩						▩			
	▩					**50625** Determine resources and quantities required	▩		▩	▩			▩			
					▩	**50630** Assign responsibilites for corrective action	▩		▩					▩		
▩	▩	▩	▩	▩	▩	**50635** Perform an impact analysis	▩						▩		▩	▩
						MEASURES OF SUCCESS										
						Are the necessary corrective actions required?						▩			▩	▩
						Is an impact study conducted for each corrective action?							▩			
						Are all the right participants participating in the corrective action?								▩		

9

Project Closure (60000)

Introduction

Sooner or later, projects come to an end. Good and bad projects share this fate. One factor separating these projects is their way of coming to a conclusion. Some projects end with the same level of professionalism as at the outset. Other projects wind down like a beginning skier descending a steep slope.

Reasons that Projects End

Projects end for many reasons. They can end after meeting all requirements. Or they can finish prematurely, owing to negative circumstances. Typically, projects conclude for one or more of these reasons:

- Changing customer requirements
- Changing market conditions
- Indecisiveness on the part of management or customer or both
- Lack of cooperation from customers
- Lack of management support
- Lack of resources (e.g., insufficient time, money people, equipment, material, supplies)
- Lack of teamwork
- Failure to meet all requirements (e.g., cost, schedule, quality)
- Politics (a shift from one "pet" project to another)
- Technical problems

60000 PROJECT CLOSURE

P² M² CYCLE

Reports	Quality Variances	New Schedule	New Budget	Network Diagram	Impact Analysis	Forms	Cost Estimates	Changed Schedule	Changed Budget	Budget Variances	Bar Chart	TASKS	Project Manager	Senior Management	Project Sponsor	Project Team	Client	Released Resources	Client Approval	Awards	Written Report	Lessons Learned	Procedures	Statistics
												RESPONSIBILITIES / **OUTPUTS**												
X	X	X	X	X	X	X	X	X	X	X	X	60100 Compile statistics	X		X								X	X
X	X	X	X	X	X	X	X	X	X			60200 Prepare Lessons Learned	X			X	X					X		
												60300 Conduct Post-Implementation Review	X	X	X		X				X			
											X	60400 Perform Winding Down Activities	X			X	X	X	X	X				
												MEASURES OF SUCCESS												
												Are statistics being compiled?											X	X
												Is a Lessons Learned Document being produced?										X		
												Will there be a post-implementation review?									X			
												Are winding down activities occurring?						X	X	X				

INPUTS

Project Requirements	Schedule	System Manuals	Project Objectives	Project Goals	Project Charter	Mission Statement	Organization Standards	Updated Reports	Updated Budget	Statistics	Statement of Work	Revised Statement of Work	Project History Files	Minutes	Management Reserve	Lessons Learned	Contingency Plan	Changed Priorities	Changed Categories	Administrative Requirements	Work Breakdown Structure	Updated Schedule	Major Participants	Management Direction	Time Estimates	Select Project Management Package	Schedule Variances	Risk Identification	Revised Schedule	Resource Histogram
																				X	X	X								
							X	X	X	X	X	X	X	X	X		X						X	X	X	X	X	X		X
		X	X	X	X	X					X	X				X														
X																						X							X	

9.1 STATISTICAL COMPILATION (60100)

As your project concludes, you want to collect as much statistical information as possible. **Statistics compilation** of cost, schedule, and quality is particularly interesting as this information allows you to compare your plans squared with reality. By spotting deviations between plans and the actual project history, you can then identify their cause. Even when plans and the actual history match, statistical analysis still allows you to analyze why something happened.

There are at least three good reasons for collecting statistics. First, statistical compilation allows you to conduct a thorough analysis of your project. Understanding the similarities and deviations between planned and actual events provides you with an excellent "barometer" of your project's overall performance. Second, you will have better management of future projects of a similar nature since you will know what went right and wrong and, consequently, will have a better idea of how to avoid mistakes. Third, you can make more accurate cost and time estimates for similar projects in the future.

60100 STATISTICAL COMPILATION

P² M² CYCLE — INPUTS / TASKS / RESPONSIBILITIES / OUTPUTS / MEASURES OF SUCCESS

TASKS	INPUTS																							RESPONSIBILITIES					OUTPUTS	
	Bar Chart	Budget Variances	Changed Budget	Changed Schedule	Cost Estimates	Forms	Impact Analysis	Network Diagram	New Budget	New Schedule	Quality Variances	Reports	Resource Histogram	Revised Schedule	Risk Identification	Schedule Variances	Select Project Management Package	Time Estimates	Management Direction	Major Participants	Updated Schedules	Work Breakdown Structure	Administrative Responsibilities	Project Manager	Senior Management	Project Sponsor	Project Team	Client	Procedures	Statistics
60105 Establish a procedure for collecting statistics																			▨				▨	▨					▨	
60110 Assign responsibilities for collecting statistics																				▨			▨	▨					▨	
60115 Determine how to collect statistics																							▨	▨					▨	
60120 Determine what subjects to collect statistics on																							▨	▨			▨		▨	
60125 Determine when to collect statistics																							▨	▨					▨	
60130 Determine tools to collect and compile statistics																					▨		▨	▨					▨	
60135 Collect and compile statistics	▨	▨	▨	▨	▨	▨	▨	▨	▨	▨	▨	▨	▨	▨	▨	▨	▨	▨				▨		▨			▨			▨
MEASURES OF SUCCESS																														
Does a procedure for compiling statistics exist?																													▨	
Are the responsibilities for compiling statistics defined?																													▨	
Is the right statistical data compiled?																														▨
Are the results of the compilations communicated to the appropriate people?																													▨	

9.2 LESSONS LEARNED (60200)

It is very important for yourself and other project managers to prepare a "**lessons learned**" document. This document is an evaluation of the project and cites successes, problems, and future opportunities.

There are two important reasons exist for preparing a lessons-learned document. One, it serves as a historical account of the project. You can use it to develop reports and provide information for audits. More importantly, the lessons-learned document allows future project managers to learn from your experience by noting what did and did not go well.

Although lessons learned vary in format, most follow an outline similar to this one:

   I. Introduction
     A. Purpose
     B. Scope
  II. Background information
 III. Executive summary
 IV. Major achievements and successes
     A. Achievements/successes
     B. Discussion of each achievement/success
  V. Major problems
     A. Discussion of each problem
     B. Recommendation for overcoming problems
 VI. Opportunities for future work

When preparing the lessons learned, you need not rely solely upon your memory. You have several sources of information to assist you. These sources include:

     Change control requests
     Cost records
     Interviews with team members, management, and customer
     Minutes from meetings
     Schedules
     Statement of work and its subsequent revisions
     Statistical compilations

The best time to prepare lessons learned is towards the end of or immediately after the conclusion of the project. Sometimes, project managers wait too long after the project, thereby lessening its benefits.

After completing the first draft, you should circulate it among key team members for their review. You may have omitted or misinterpreted something. Once you complete the final draft, give everyone a copy, especially other project managers and your management.

Occasionally, the lessons learned fail to prove beneficial. This can be attributable to several reasons. First, some project managers are not thorough enough to produce a meaningful document. They may fear repercussions for writing the "wrong thing"—something that could later haunt them. Second, they may turn the document into a political essay—they include only comments and recommendations that make them look good. Third, they may use the document to point fingers—they blame someone else for the problems of the project. Finally, they may make the document inaccessible to the people who need to read it, such as management, team members, and other project managers. When this happens, future project managers often make the same mistakes.

The best lessons learned are written by project managers who do not fear repercussions from what they write. They know to write an honest appraisal that others can have access to and benefit from

60200 LESSONS LEARNED

Cost Estimates	Contingency Plan	Changed Schedule	Changed Budget	Changed Priorities	Changed Categories	Budget Variances	Bar Chart	P2 M2 CYCLE / TASKS	Project Manager	Senior Management	Project Sponsor	Project Team	Client	Lessons Learned
								TASKS						
								60205 Assign responsibilities for Lessons Learned	▓			▓	▓	▓
								60210 Develop an outline for Lessons Learned	▓					▓
▓	▓	▓	▓	▓	▓	▓	▓	**60215 Determine sources of information**	▓		▓			▓
▓	▓	▓	▓	▓	▓	▓	▓	**60220 Identify suggestions for improvement**	▓			▓	▓	▓
								60225 Prepare Lessons Learned	▓					▓
								MEASURES OF SUCCESS						
								Are the appropriate participants providing input to the Lessons Learned?						▓
								Is the Lessons Learned agreed upon by all major participants?						▓
								Are the responsibilities for creating, reviewing, approving, and distributing the						▓
								Does the Lessons Learned have the appropriate level of detail?						▓
								Is the Lessons Learned available to all project managers of similar projects?						▓
								Is the Lessons Learned objective?						▓

	Management Direction	Organization Standards	Updated Schedule	Updated Reports	Updated Budget	Time Estimates	Statistics	Statement of Work	Selected Project Manag	Schedule Variances	Risk Identification	Revised Statement of	Resource Histograms	Reports	Quality Variances	Major Participants	Project History Files	New Schedule	New Budget	Network Diagram	Minutes	Management Reserve	Impact Analysis	Forms
																								INPUTS
	▨																							
		▨																						
	▨	▨	▨	▨	▨	▨	▨	▨	▨	▨	▨	▨	▨	▨	▨	▨	▨	▨	▨	▨	▨	▨	▨	▨
	▨	▨	▨	▨	▨	▨	▨	▨	▨	▨	▨	▨	▨	▨	▨	▨	▨	▨	▨	▨	▨	▨	▨	▨
	▨	▨																						

9.3 POST-IMPLEMENTATION REVIEW (60300)

After the new product is in the hands of the client, it is good practice to institute a **post-implementation review**. Basically, this involves investigating the technical and business performance of the product and the sustaining, or maintenance, activities that will be supporting the client during the use of the product.

The post-implementation review entails these activities:

> Determining whether goals and objectives of the project have been achieved
> Ascertaining whether the product has met the needs of the client
> Assessing whether the benefits of the product have been realized
> Determining the customer's overall satisfaction with the system
> Identifying areas of satisfactory performance and areas for improvement and
> Reviewing sustaining support, paying special attention to the types of changes, their impact, and how they are handled

When performing the post-implementation review, consider these points:

1. Utilize independent individuals to conduct the review. They should not be former members of the project team or members from the client's community, as they will have a vested interest in the outcome of the review and the review, no matter how well-intentioned, will appear subjective.

2. Identify the scope of the review before starting. The review must have a definable scope. Without defined limits, it can easily turn into a review that oversteps acceptable boundaries and causes political havoc, especially when the project has involved a large number of players. Egos and careers can easily be viewed by some as being at stake.

3. Wait three to six months before starting a review. Of course, the time needed to do the review depends on the complexity and size of the project.

4. Don't be shy about conducting the review. Use all available means for collecting information. These include documentation, interviews, and tests to determine whether goals, objectives, standards, and requirements have been met.

5. Publish a report of the findings. This report should be available not only to the supporting organization and the client but also to project managers working on projects of similar or related products. Your findings will help them avoid the same pitfalls and problems so that their projects finish on time and within budget while delivering a product of high quality.

6. Review documentation produced during the earlier and latter parts of the project to determine the criteria to conduct the review. Documentation should include looking at the statement of work, cost/benefit studies, requirements documentation, change control records, test plans, and test results. Such documentation gives a good idea of the business and technical goals, objectives, and requirements needed to conduct the review.

60300 POST-IMPLEMENTATION REVIEW

P² M² CYCLE

INPUTS → TASKS → RESPONSIBILITIES → OUTPUTS → MEASURES OF SUCCESS

TASKS

TASKS	Schedule	Management Direction	System Manuals	Statement of Work	Revised Statement of Work	Project Objectives	Project Goals	Project Charter	Mission Statement	Lessons Learned	Project Manager	Senior Management	Project Sponsor	Project Team	Client	Written Report
	INPUTS										RESPONSIBILITIES					OUTPUTS
60305 Determine best time to conduct reviews	▓	▓									▓					▓
60310 Contact reviewers		▓									▓					▓
60315 Assemble material for reviews		▓	▓	▓	▓	▓	▓	▓	▓	▓	▓					▓
60320 Conduct post-implementation review			▓	▓	▓	▓	▓	▓	▓	▓	▓	▓			▓	▓

MEASURES OF SUCCESS

MEASURES OF SUCCESS	Written Report
Is the post-implementation review occurring within a reasonable timeframe?	▓
Are the reviewers independent?	▓
Is the report complete?	▓
Is the report distributed to the right people?	▓

9.4 WINDING DOWN ACTIVITIES (60400)

Collecting statistics and preparing lessons learned are not the only actions to take as your project concludes. You must perform other, no less important, tasks.

One task is managing personnel. As activities on a project decline in number and magnitude, you need to release or reassign people.

You may elect to release people for two reasons. Idle personnel not only increase labor costs but can interfere with the activities of team members who are busy.

You may also elect to reassign people to help other team members complete tasks. Although intuitively this should increase productivity, reassignment can sometimes have the opposite effect. Reassigned personnel may require training and there may be the need for more communications among team members.

Another responsibility is to recognize those participants who have performed above the norm—this can include the entire team or just certain individuals.

Recognition gives people a sense of accomplishment and a feeling of having done the job well. The problem with bestowing recognition is deciding how to do it. While money is a powerful motivator, not everyone is motivated the same way by it; besides, most project managers lack the power to grant monetary rewards. Still, you can give recognition in many ways. You can prepare letters of appreciation and commendation to be placed in the personnel file. You can give deserving employees certificates and plaques accompanied by cash awards, or send them to a training program, or even reward them with dinner at a nice restaurant. The idea is to give people a feeling that they are significant participants.

One last step is to obtain formal approval of the final product from the customer. You can obtain signatures on a document, which will serve as a formal record of the client's approval. By receiving the client's approval, you reduce the chance of being blamed for anything concerning the final product that displeases the customer. This helps you share responsibility for the outcome of the project.

60400 WINDING DOWN ACTIVITIES

INPUTS				P² M² CYCLE	RESPONSIBILITIES					OUTPUTS		
Project Requirements	Updated Schedule	Revised Schedule	Changed Schedule		Project Manager	Senior Management	Project Sponsor	Project Team	Client	Released Resources	Client Approval	Awards
				TASKS								
■				**60405** Obtain approval of final product from client	■				■		■	
	■	■	■	**60410** Recognize participants performing above the norm	■		■					■
	■	■	■	**60415** Release or reassign people	■		■			■		
	■	■	■	**60420** Release or reassign equipment, supplies, materials, and facilities	■		■			■		
				MEASURES OF SUCCESS								
				Does the client approve of the product?							■	
				Are the top performers recognized for their contributions?								■
				Is resource idleness kept to a minimum?						■		
				Are efforts taken to keep morale high up to achieving the final milestone?						■		

10

Keys to Success

Regardless of industry or position within a company, all project managers perform six fundamental functions, with varying levels of success: leading, defining, planning, organizing, controlling, and closing. The Practical Project Management Methodology, provides tools, techniques, and knowledge for successful project managers to apply when performing these six functions.

As **Exhibit 10.0-1** shows, effective project management involves a wide spectrum of activities. You need to define project management goals and objec-

---

Effective Project Management = Statement of Work
+ Project Announcement
+ Work Breakdown Structure
+ Project Team
+ Project Documentation
+ Administrative Activities
+ Time Estimates
+ Budgets
+ Quality Standards
+ Management Skills
+ Project Termination Activities
+ Motivation

THE FORMULA FOR EFFECTIVE PROJECT MANAGEMENT
EXHIBIT 10-1

tives, acquire management support, identify all the tasks to build the product, assemble a team, develop a good administrative apparatus, estimate times to perform each task and the entire project, allocate funds and available resources, establish acceptable levels of quality, acquire the necessary managerial and technical skills, conclude the project efficiently, and, motivate everyone throughout the entire project.

These tasks can be accomplished in an orderly manner. As **Exhibit 10-2** illustrates, you can perform them in a logical sequence to implement the tools and techniques of project management.

You can use the chart to help you make project management a reality for your project. To understand each step in the process, just follow the instructions on the chart:

In the end, it is up to you to decide how to implement the tools, techniques, and resources of project management. As the project manager, you decide the sequence of actions and the level of detail required. In some cases, you may find that a memo of understanding is all you need for a statement of work; one estimate for each activity rather than three; a bar chart rather than a network diagram; or hand-drawn rather than computer-drafted schedules.

Only you can make these decisions. More often than not, project managers must decide the degree of project management disciplines to implement. That is the way it should be, too. You know the project better than anyone else and play a critical role in its outcome. Besides, you were selected as project manager because you have demonstrated an ability to make sound judgments. Being able to select the tools, techniques, and resources to implement and knowing to what degree to perform the six fundamental functions is all part of the judgment you need to be an effective project manager. The responsibility is all yours.

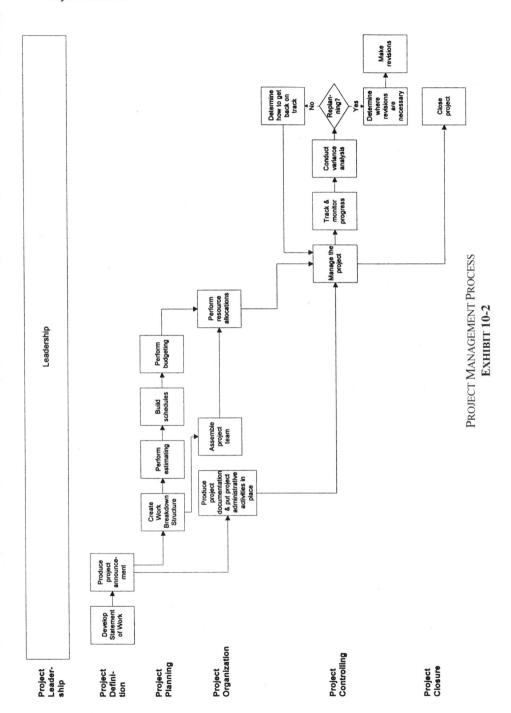

PROJECT MANAGEMENT PROCESS
EXHIBIT 10-2

Glossary

Action item log A medium to capture information about unforeseen circumstances.

Activity description form A document used to capture the basic information for constructing network diagrams.

Activity estimating form A medium to calculate and record estimates for each activity and an entire project.

Actual costs The monies spent after managing part or all of a project.

Actual resource utilization costs form A medium to record the actual costs associated with using resources.

Ad hoc Group(s), such as a project steering committee advisory group or quality assurance, that may be involved with the project at the discretion of the project sponsor, project manager, or senior management.

Ad hoc meetings Group sessions held whenever a need arises.

Arrow diagramming method A network diagramming technique using circles and arrows to show relationships among tasks. It is used most often in the construction environment.

Change agent A person or organization that plans, organizes, and executes a change.

Change board A group consisting of a project manager, team leader, project sponsor, and customer representatives.

Change control Policies and procedures established to detect, analyze, evaluate, and implement modifications to all baselines in your project.

Change control log A medium to record request alterations to processes and deliverables.

Change control procedure A document showing how to manage changes.

Change management Introducing change in a manner that minimizes negative impacts on productivity.

Change request form A medium to record an inquiry about modifying a component of the product being built.

Change sponsor A person(s) or organizations(s) that provides the "political" muscle to effect a change.

Change target A person(s), place(s), or organization(s) that is the object of a change.

Checkpoint review meetings Group sessions that occur whenever a project finishes a certain phase, task, or milestone.

Client The person(s) or organization(s) for whom the product is built.

Client review committee A group of individuals assigned to review project deliverables and to provide client acceptance of the project.

Closing The fifth of six functions for managing a project. Closing involves compiling data, converting the data into information, and providing a smooth transition from product development to implementation.

Complete project cost form A medium used to capture the costs for a task or an entire project.

Contingency plan form A medium used to determine anticipated responses to unexpected circumstances.

Contingency planning Anticipating responses to circumstances that can negatively impact a project.

Controlling The fourth of six functions for managing a project. Controlling involves ensuring that projects proceed according to plans and in a manner that maximizes the effectiveness of an organizational structure.

Corrective action A decision of project managers about whether to follow their plans or not and if not, develop, and implement new ones.

Corrective change A fix to a problem that was overlooked sometime during the project.

Cost variance The difference between budgeted or estimated costs and actual costs incurred for each activity and for the entire project.

Critical path The activities in the network diagram that cannot slide at any moment in time without impacting subsequent activities and, ultimately, the project completion date.

Critical path method (CPM) A scheduling technique used to identify the critical activities to completion.

Current schedule A network diagram or bar chart that reflects the target schedule and the status against the items contained therein.

Decision A work flow symbol indicating a choice between two or more paths.

Defining The first of six major functions for managing a project. Defining involves identifying at a high level the who's, what's, when's, where's, and how's of a project.

Delay A work flow symbol reflecting the time period elapsing when no action occurs.

Document A work flow symbol representing a document that is being processed.

Eighty-hour rule Also known as the two-week rule. If a task breakdown structure (TBS) element requires more than two weeks of work, or 80 working hours, then explode the TBS down another level.

Employee assignment form A medium to record what resources are used on which tasks.

Estimate-at-completion (EAC) Cost data expected to spend on each activity and an entire project.

Estimate costs The monies required to manage a project.

Estimated labor usage form A medium to calculate estimates regarding labor costs.

Estimated nonlabor usage form A medium to calculate estimates regarding nonlabor costs.

Estimating form A document used to record time estimates.

Float The time an activity can slide without fear of impacting subsequent activities and, ultimately, the project completion date.

Form A medium used to capture, store, and transport data to the appropriate destination.

Global efficiency factor An estimating technique that assumes employees are 100 percent productive and then deducts from that 100 percent an estimated percentage for each deficiency identified for the project.

Goals Meaningful statements describing what the project will achieve.

Item-by-item format The outlay of a procedure that describes a series of topics related only by general subject matter.

Leadership The concept of influencing people in the attainment of an organization's goals and objectives.

Leading The sixth of six functions for managing a project. Leading involves creating an environment that encourages the best in people and in a way that meets project goals and objectives.

Lessons learned An evaluation document produced after a project, citing successes, problems, and future opportunities.

Leveled histogram A chart profiling a smoother employment of resources after modifying the schedule.

Maintenance change An alteration made to keep something current.

Major change A dramatic alteration to the schedule, product baselines, budget, or anything else deemed critical to the project.

Major priority A change assigned that needs immediate attention, otherwise, it will stop the project.

Management by confusion A management style characterized by random or circular steps.

Management by crisis A management style in which managers find themselves constantly reacting rather than proacting to their circumstances.

Management by drives A management style characterized by nothing happening for a long time, followed by a mad rush at the last minute.

Management by efficiency and effectiveness A management style characterized by smooth handling of crises and constant, steady action.

Management reserve A percentage of the estimate to complete a project that is set aside to adjust to circumstances when actual costs exceed budget.

Matrix structure A group of individuals supporting multiple projects while working for a functional manager.

Meetings Group sessions that should result in accomplishing goals and objectives efficiently and effectively.

Minor change An alteration that is not a major change and that does not ultimately affect the outcome of the project.

Minor priority A change assigned that does not require immediate attention but that must be confronted eventually to complete the project.

Motivation Encouragement of people to participate actively to attain project goals and objectives.

Narrative format The outlay of a procedure that describes the "what" and "how to" information in essay style.

Network diagrams A type of schedule showing the start and stop dates for tasks and the relationships among tasks.

No priority A change assigned that can be addressed if time permits.

Objectives Measurable subsets of goals.

Organization The third of six major functions for managing a project. Organization involves instituting a structure to maximize the efficiency and effectiveness of a project.

Overrun A situation in which the difference between the budget-to-date and the actuals-to-date is more than expected.

PERT estimate A technique using three variables for determining the time required to complete a task or an entire project. The three variable are usually most likely estimate, most optimistic estimate, and most pessimistic estimate.

Planning The second of six major functions for managing a project. Planning involves determining the tasks required to complete goals and objectives, the most significant tasks, and the resources required to complete the project.

Post-implementation review Investigation of the technical and business performance of the product and the sustaining, or maintenance, activities supporting the client during the use of the product.

Practical project management methodology An approach to project management that stresses flexibility, results, participation, accountability, minimal hierarchy, less documentation, and continuous improvement.

Precedence diagramming method A network diagram technique that enables different logical relationships among tasks. It is often used in nonconstruction environments, such as service environments.

Predecessor An activity that completes before another one can start.

Predecessor-successor schedule report A medium detailing the schedule progress made on tasks and their relationships to one another.

Proactive project management A category of management style characterized by managers handling their projects efficiently and effectively.

Problem occurrence record form A medium to record negative situations.

Process A work flow symbol representing a function.

Product breakdown structure (PBS) The first component of the work breakdown structure that shows a decomposition of the final product.

Productivity adjustment percent An estimating technique similar to the global efficiency factor; however, the percentage used is based upon objective measurement.

Program evaluation and review technique (PERT) A scheduling technique used to determine the approximate time to complete tasks.

Project An endeavor that begins at a specific point in time, must complete some time in the future, and deliver a final product.

Project announcement A brief, one-page memo announcing the existence of a project.

Project closure The last phase of a project.

Project controlling Applying ways to ensure that projects proceed according to plans.

Project cost report A medium detailing the monetary progress made on tasks and for the entire project.

Project history files A compilation of information about a project.

Project library A place to store project documentation to protect it and make it accessible to everyone.

Project management software Application packages used to manage projects. Applications include scheduling, word processing, spreadsheets, graphics, database management, and communications.

Project manager The person(s) responsible and accountable for the outcome of a project.

Project manual A compilation of useful information that project participants can refer to whenever they have a question.

Project office A central location for the resources of the project and that serves as a "command and control center" for the project.

Project sponsor The person(s) project managers report to and the ones who make major decisions regarding the fate of projects.

Project status report form A medium to record an assessment of how well a project is progressing at specific intervals.

Project steering committee A group of individuals that provides direction to the project team.

Project team The group of individuals who support the project manager by working together efficiently and effectively to deliver a product that satisfies the client's needs.

Reactive project management A management style characterized by managers handling their projects in a helter skelter manner.

Replanning Restructuring part or all of a project.

Report A medium to display information about progress regarding schedule, budget, and quality.

Resource allocation Assignment of people, supplies, materials, etc. for the completion of tasks.

Resource cost estimate form A medium for recording estimated costs.

Resource histogram A chart profiling how well project staff is used, given the current schedule.

Resource histogram form A document used to record information necessary for building a resource histogram.

Resource leveling The process of optimizing the use of resources to avoid having a histogram riddled with peaks and valleys.

Responsibility matrix A table showing the tasks to be performed on one axis and the people to perform those tasks on the other.

Risk analysis Identification of the critical activities of a project and the threats to them.

Risk assessment Determination of the probability of occurrence of a threat and the economic, operational, and technical impact on an organization.

Risk assessment checklist A list for identifying risks to a project.

Risk control Recognition of the threats to a project and development of plans to minimize their occurrence.

Risk control log A medium used to develop contingency plans for dealing with each threat to a critical activity.

Risk management Determination of the measures to take to prevent a threatened negative consequence from occurring or to lessen its impact.

Schedules Diagrams that identify tasks, apply time estimates for those tasks, and create start and stop dates for each task and the entire project.

Schedule variance The difference between planned and actual start and stop dates.

Senior management The person(s) responsible for determining which projects will be initiated and for assigning a project sponsor to each project.

Sequential format The outlay of a procedure that describes the "what" and "how to" information in outline style.

Staff meetings Weekly or biweekly meetings held for team members to deliver announcements or exchange information concerning the project.

Statement of work (SOW) A document defining the scope of the project and the responsibilities of the participants.

Statistical compilation Collection of data on project performance and conversion of that data into information.

Status assessment Analysis of a project's progress in terms of schedule, budget, and quality.

Status collection Employment of measures to determine how well certain activities and the entire project are progressing.

Status review meetings Group sessions that collect and assess status on a regular basis to ascertain progress.

Storage A work flow symbol representing paper or information stored someplace.

Successor An activity that starts after a preceding one completes.

Support The task of doing whatever is necessary to help protect participants perform their tasks efficiently and effectively.

Target schedule Also known as baseline schedule. The target schedule is an agreed-upon network diagram or bar chart.

Task breakdown structure (TBS) The second component of the WBS, comprising tasks, subtasks, and so on that are required to build each subproduct and ultimately contribute to the building of the overall product.

Task force A group of individuals working full-time on a project. Their attention is devoted entirely towards achieving the overall project goal.

Time estimates The amount of effort required to complete tasks and subtasks.

Traditional project management methodologies Approaches to managing projects that stress predictability; mathematics; centralization; functionality; change resistance; efficiency; and focus on cost, schedule, and quality.

Underrun The situation in which the difference between the budget-to-date and the actuals-to-date is less than expected.

Unscientific estimate Also known as a scientific wildly assumed guess (SWAG). The unscientific estimate is a technique for determining the time required to complete a task or an entire project.

Variance The difference between planned and actual performance.

Vector A symbol representing the direction of logic throughout a work flow.

Vision An idea of what the project hopes to achieve.

Winding down activities Managing resources cost-effectively as the project nears completion.

Work breakdown structure (WBS) A hierarchical listing of the products, sub-products, tasks, and subtasks required to complete the project.

Work flows A medium for graphically displaying the contents of procedures.

Work package level The lowest level elements shown on each leg within the WBS.

Appendix: Instructions for Using the Disk with the Files for Frameworks, Forms, and Microsoft Project™

Extracting P^{2}M^2 Files

The diskette accompanying this book contains a set of files for the frames and forms (compatible with Microsoft Excel version 5.0 or higher) along with a generic project plan (compatible with Microsoft Project version 3.0 or higher). The files on the diskette are compressed and need to be expanded and copied to your hard drive.

For Windows 95 or Windows NT:
- Start Windows
- Insert the diskette in drive a
- Click the Start button (File Menu in Windows NT)
- Click Run
- Type a:\p2m2 and click OK

For Windows version 3.x:
- Start Windows
- Insert the diskette in drive a
- Choose Run from the File menu
- Type a:\p2m2 and choose OK

You can then open the files from the p2m2 directory on your hard drive.

Frameworks and Forms

After extracting the P^2M^2 files from the diskette accompanying this book, you will have all of the frameworks on your hard drive, in the directory C:\P2M2\FRAMES and all of the forms on your hard drive, in the directory C:\P2M2\FORMS. The frameworks and forms are compatible with Microsoft Excel version 5.0 or higher.

The frameworks and forms all have the .xls extension and can be used as they appear or be modified to reflect your particular circumstances.

The frameworks are listed below:

File	Title	File	Title
00000	Project Environment Assessment	40200	Team Organization
		40300	Project Procedures
10000	Project Leadership	40400	Project History Files
10100	Vision	40500	Work Flows
10200	Communication	40600	Forms
10300	Motivation	40700	Reports
10400	Direction	41000	Newsletters
10500	Support	41100	Project Office
10600	Team Building	41200	Project Manual
20000	Project Definition	50000	Project Control
20100	Goals and Objectives Determination	40800	Project Library
		40900	Memos
20200	Statement of Work	50100	Status Collection and Assessment
20300	Project Announcement		
20400	Project Launch	50200	Contingency Planning
20500	Roles and Responsibilities	50300	Meetings
30000	Project Planning	50400	Change Control
30100	Work Breakdown Structure	50500	Replanning
30200	Estimating	50600	Corrective Action
30300	Risk Control	60000	Project Closure
30400	Schedules	60100	Statistical Compilation
30500	Resources Allocation	60200	Lessons Learned
30600	Cost Calculation	60300	Post-Implementation Review
40000	Project Organization	60400	Winding Down Activities
40100	Automated Project Management		

The forms are listed below:

File	Title	File	Title
Actdes	Activity Description Form	Predsucc	Predecessor-Successor Report
Actest	Activity Estimating Form		
Actitlog	Action Item Log	Proboccr	Problem Occurrence Record Form
Chngcont	Change Control Log		
Chngreq	Change Request Form	Probrep	Problem Report
Compcost	Complete Project Cost Form	Projcost	Project Cost Report
		Projstat	Project Status Report
Conting	Contingency Plan Form	Rescost	Resource Cost Estimate Form
Emassign	Employee Assignment Form		
Estform	Estimating Form	Reshisto	Resource Histogram Form
Labusag	Estimated Labor Usage Form	Resutil	Actual Resource Utilization Costs Form
Nonlabus	Estimated Non-Labor Usage Form	Riskcon	Risk Control Form
		Sow	Statement of Work

Just follow these steps to get started:

1. Start Windows 3.x or Windows 95.
2. Run Microsoft Excel.
3. Select the P2M2 directory on your C drive.
4. Open the "Frames" or the "Forms" subdirectory.
5. Select the desired file.
6. Use or modify the file as desired.*
7. Save the file under a different filename.

Microsoft Project™

After extracting the P2M2 files from the diskette accompanying this book, you will have the generic project plan on your hard drive, in the directory C:\P2M2\MPP. The generic project plan is compatible with Microsoft Project version 3.0 or higher.

It contains the (generic) work breakdown structure and network diagram based upon the contents of your book. Each of the approximately 350 entries has a unique number and description (pp. 265 and 266), such as "40300 Project proce-

---

*To prevent overwriting the original file, it is advisable to save your new file under a different name before making modifications. (This is accomplished by using "Save As" under the "File" menu.)

dures," for sub-products and for their lower tasks, such as "40305 Determine type of procedures" and "40310 Determine who should prepare the procedures." The numbers and descriptions match the ones appearing in the frameworks of the methodology.

The following apply to each of the work package level items:

- Predecessor–successor relationships
- A finish-to-start type of relationship
- Lags of 0 with all preceding and succeeding tasks
- Generic assignments (for 8-hour work days) as they appear in the frameworks
- Duration of 1 day

When using this file, you can modify the contents to fit the particular needs of your project. Here are some examples:

- Add, modify, or remove tasks
- Change task descriptions
- Remove or change task dependencies, including predecessor–successor relationships, lag, and relationship type
- Replace generic assignment names with real names
- Link the network diagram to another one for a project, such as for business process reengineering, data warehousing, document management, intranet/Internet projects as well as for nontechnical projects, such as marketing campaigns, publishing documentation, and course development projects.

Just follow these steps to get started:

1. Start Windows 3.x or Windows 95.
2. Run Microsoft Project (Version 3.0 or higher).
3. Select the P2M2 directory on your C drive..
4. Open the MPP subdirectory.
5. Open the file PM-METH4.MPP.
6. Use or modify the file as desired.*
7. Save the file before closing Microsoft Project. Use a different filename, than the original provided with your book, if you haven't done so already.
8. Use the new file to manage your project.

Note: The timescale associated with the initial task (00000, Project Assessment) begins on January 15, 1995. You will need to scroll your timescale back to that date in order to see the schedule flows.

---

*To prevent overwriting the original file, it is advisable to save your new file under a different name before making modifications. (This is accomplished by using "Save As" under the "File" menu.)

		Practical Project Management Methodology					

ID	Name			Duration	Scheduled Start	Scheduled Finish	Predecessors
2	00005 Determine which projects in			1d	1/16/95 8:00am	1/16/95 5:00pm	

ID	Resource Name	Units	Work	Delay	Scheduled Start	Scheduled Finish	
2	Project manager	1	8h	0h	1/16/95 8:00am	1/16/95 5:00pm	

ID	Name			Duration	Scheduled Start	Scheduled Finish	Predecessors
3	00010 List the major pressures co			1d	1/17/95 8:00am	1/17/95 5:00pm	2

ID	Predecessor Name					Type	Lag
2	00005 Determine which projects in your environment are examples of management b					FS	0d

ID	Resource Name	Units	Work	Delay	Scheduled Start	Scheduled Finish	
2	Project manager	1	8h	0h	1/17/95 8:00am	1/17/95 5:00pm	

ID	Name			Duration	Scheduled Start	Scheduled Finish	Predecessors
4	00015 Determine whether your co			1d	1/18/95 8:00am	1/18/95 5:00pm	3

ID	Predecessor Name					Type	Lag
3	00010 List the major pressures confronting projects in general and project mana					FS	0d

ID	Resource Name	Units	Work	Delay	Scheduled Start	Scheduled Finish	
2	Project manager	1	8h	0h	1/18/95 8:00am	1/18/95 5:00pm	

ID	Name			Duration	Scheduled Start	Scheduled Finish	Predecessors
5	00020 List some major problems c			1d	1/19/95 8:00am	1/19/95 5:00pm	4

ID	Predecessor Name					Type	Lag
4	00015 Determine whether your company generally follows a proactive or reactive					FS	0d

ID	Resource Name	Units	Work	Delay	Scheduled Start	Scheduled Finish	
2	Project manager	1	8h	0h	1/19/95 8:00am	1/19/95 5:00pm	

ID	Name			Duration	Scheduled Start	Scheduled Finish	Predecessors
6	00025 Assess how effectively pro			1d	1/20/95 8:00am	1/20/95 5:00pm	5

ID	Predecessor Name					Type	Lag
5	00020 List some major problems confronting project managers in your company					FS	0d

ID	Resource Name	Units	Work	Delay	Scheduled Start	Scheduled Finish	
2	Project manager	1	8h	0h	1/20/95 8:00am	1/20/95 5:00pm	

ID	Name			Duration	Scheduled Start	Scheduled Finish	Predecessors
7	00030 List the major participants c			1d	1/17/95 8:00am	1/17/95 5:00pm	2

ID	Predecessor Name					Type	Lag
2	00005 Determine which projects in your environment are examples of management b					FS	0d

ID	Resource Name	Units	Work	Delay	Scheduled Start	Scheduled Finish	
2	Project manager	1	8h	0h	1/17/95 8:00am	1/17/95 5:00pm	

ID	Name			Duration	Scheduled Start	Scheduled Finish	Predecessors
8	00035 List the major internal polic			1d	1/18/95 8:00am	1/18/95 5:00pm	7

ID	Predecessor Name			Type	Lag		
7	00030 List the major participants on your project			FS	0d		

ID	Resource Name	Units	Work	Delay	Scheduled Start	Scheduled Finish	
2	Project manager	1	8h	0h	1/18/95 8:00am	1/18/95 5:00pm	

ID	Name			Duration	Scheduled Start	Scheduled Finish	Predecessors
9	00040 Determine the external acto			1d	1/19/95 8:00am	1/19/95 5:00pm	8

ID	Predecessor Name					Type	Lag
8	00035 List the major internal policies and procedures that affect your project					FS	0d

ID	Resource Name	Units	Work	Delay	Scheduled Start	Scheduled Finish	
2	Project manager	1	8h	0h	1/19/95 8:00am	1/19/95 5:00pm	

ID	Name			Duration	Scheduled Start	Scheduled Finish	Predecessors
10	00045 List the forces or circumsta			1d	1/20/95 8:00am	1/20/95 5:00pm	9

ID	Predecessor Name			Type	Lag		
9	00040 Determine the external actors that you must deal with			FS	0d		

ID	Resource Name	Units	Work	Delay	Scheduled Start	Scheduled Finish	
2	Project manager	1	8h	0h	1/20/95 8:00am	1/20/95 5:00pm	

ID	Name			Duration	Scheduled Start	Scheduled Finish	Predecessors
11	00050 Develop a historical overvie			1d	1/23/95 8:00am	1/23/95 5:00pm	6

ID	Predecessor Name					Type	Lag
6	00025 Assess how effectively project managers are performing the four fundament					FS	0d

ID	Resource Name	Units	Work	Delay	Scheduled Start	Scheduled Finish	
2	Project manager	1	8h	0h	1/23/95 8:00am	1/23/95 5:00pm	

ID	Name			Duration	Scheduled Start	Scheduled Finish	Predecessors
12	00055 Determine what environme			1d	1/24/95 8:00am	1/24/95 5:00pm	11

ID	Predecessor Name			Type	Lag		
11	00050 Develop a historical overview of project management in your company			FS	0d		

ID	Resource Name	Units	Work	Delay	Scheduled Start	Scheduled Finish	
2	Project manager	1	8h	0h	1/24/95 8:00am	1/24/95 5:00pm	

SAMPLE LISTING OF THE 350 ENTRIES
IN THE NETWORK DIAGRAM

ID	Name	Duration	Scheduled Start	Scheduled Finish	Predecessors
13	00060 List the obstacles you or yc	1d	1/25/95 8:00am	1/25/95 5:00pm	10,12

ID	Predecessor Name		Type	Lag
10	00045 List the forces or circumstances that could cause disequilibrium		FS	0d
12	00055 Determine what environments within your company could best use project ma		FS	0d

ID	Resource Name	Units	Work	Delay	Scheduled Start	Scheduled Finish
2	Project manager	1	8h	0h	1/25/95 8:00am	1/25/95 5:00pm

ID	Name	Duration	Scheduled Start	Scheduled Finish	Predecessors
14	00065 Determine ways to overcon	1d	1/26/95 8:00am	1/26/95 5:00pm	13

ID	Predecessor Name		Type	Lag
13	00060 List the obstacles you or your management could face when implementing pr		FS	0d

ID	Resource Name	Units	Work	Delay	Scheduled Start	Scheduled Finish
2	Project manager	1	8h	0h	1/26/95 8:00am	1/26/95 5:00pm

ID	Name	Duration	Scheduled Start	Scheduled Finish	Predecessors
17	10105 Review the assessment of	1d	1/27/95 8:00am	1/27/95 5:00pm	14

ID	Predecessor Name	Type	Lag
14	00065 Determine ways to overcome these obstacles	FS	0d

ID	Resource Name	Units	Work	Delay	Scheduled Start	Scheduled Finish
2	Project manager	1	8h	0h	1/27/95 8:00am	1/27/95 5:00pm

ID	Name	Duration	Scheduled Start	Scheduled Finish	Predecessors
18	10110 Consult with project spons	1d	1/30/95 8:00am	1/30/95 5:00pm	17

ID	Predecessor Name	Type	Lag
17	10105 Review the assessment of the project environment	FS	0d

ID	Resource Name	Units	Work	Delay	Scheduled Start	Scheduled Finish
2	Project manager	1	8h	0h	1/30/95 8:00am	1/30/95 5:00pm
4	Project sponsor	1	8h	0h	1/30/95 8:00am	1/30/95 5:00pm

ID	Name	Duration	Scheduled Start	Scheduled Finish	Predecessors
19	10115 Prepare the project vision	1d	1/31/95 8:00am	1/31/95 5:00pm	18,20

ID	Predecessor Name	Type	Lag
18	10110 Consult with project sponsor	FS	0d
20	10120 Consult with major project participants	FS	0d

ID	Resource Name	Units	Work	Delay	Scheduled Start	Scheduled Finish
2	Project manager	1	8h	0h	1/31/95 8:00am	1/31/95 5:00pm

ID	Name	Duration	Scheduled Start	Scheduled Finish	Predecessors
20	10120 Consult with major project	1d	1/30/95 8:00am	1/30/95 5:00pm	17

ID	Predecessor Name	Type	Lag
17	10105 Review the assessment of the project environment	FS	0d

ID	Resource Name	Units	Work	Delay	Scheduled Start	Scheduled Finish
2	Project manager	1	8h	0h	1/30/95 8:00am	1/30/95 5:00pm
3	Senior management	1	8h	0h	1/30/95 8:00am	1/30/95 5:00pm

ID	Name	Duration	Scheduled Start	Scheduled Finish	Predecessors
21	10125 Distribute the vision	1d	2/1/95 8:00am	2/1/95 5:00pm	19

ID	Predecessor Name	Type	Lag
19	10115 Prepare the project vision	FS	0d

ID	Resource Name	Units	Work	Delay	Scheduled Start	Scheduled Finish
2	Project manager	1	8h	0h	2/1/95 8:00am	2/1/95 5:00pm
3	Senior management	1	8h	0h	2/1/95 8:00am	2/1/95 5:00pm
4	Project sponsor	1	8h	0h	2/1/95 8:00am	2/1/95 5:00pm
5	Project team	1	8h	0h	2/1/95 8:00am	2/1/95 5:00pm
6	Client	1	8h	0h	2/1/95 8:00am	2/1/95 5:00pm

ID	Name	Duration	Scheduled Start	Scheduled Finish	Predecessors
23	10205 Develop a project commun	1d	2/2/95 8:00am	2/2/95 5:00pm	21

ID	Predecessor Name	Type	Lag
21	10125 Distribute the vision	FS	0d

ID	Resource Name	Units	Work	Delay	Scheduled Start	Scheduled Finish
2	Project manager	1	8h	0h	2/2/95 8:00am	2/2/95 5:00pm

ID	Name	Duration	Scheduled Start	Scheduled Finish	Predecessors
24	10210 Document the plan	1d	2/3/95 8:00am	2/3/95 5:00pm	23

ID	Predecessor Name	Type	Lag
23	10205 Develop a project communications plan	FS	0d

ID	Resource Name	Units	Work	Delay	Scheduled Start	Scheduled Finish
2	Project manager	1	8h	0h	2/3/95 8:00am	2/3/95 5:00pm

SAMPLE LISTING OF THE 350 ENTRIES
IN THE NETWORK DIAGRAM (CONT.)

Index